*Active Leadership in Education
Enterprise and Engagement*

The
Safety Audit

The
Safety Audit
Designing Effective
Strategies

ROGER SAUNDERS

FINANCIAL TIMES

PITMAN PUBLISHING

TO ISABELLE

Pitman Publishing
128 Long Acre, London WC2E 9AN

A Division of Longman Group UK Limited

First published in 1992
Reprinted 1994

© Roger Saunders 1992

British Library Cataloguing in Publication Data
A CIP catalogue record for this book can be obtained
from the British Library

ISBN 0 273 03448 0

Typeset by PanTek Arts
Printed and bound in Great Britain by
Biddles Ltd, Guildford and King's Lynn

CONTENTS

FOREWORD

I was delighted to be asked to write something about this book. I have been involved with the process of safety auditing for over 30 years and pleased that at last some general advice about the measurement of safety performance has been written for the manager. Such a book as this, which covers the four P's involved in safety auditing, covers in sufficient detail the safety examination of policy, procedures, practice and programmes. A book such as this will need a much wider audience than just the manager and would be invaluable to personnel managers and, of course, trained safety practitioners trained to the competency of the government-backed National Council of Vocational Qualifications.

The book begins by clarifying what exactly safety auditing is and what its objectives must be. From this it is possible to consider tasks and role analysis and on the very subject of safety performance management itself. Chapters deal with the four principal areas involved in safety auditing and cover the safety audit on safety policy, procedures, practice and programmes. Additional chapters look at safety audit documentation together with three chapters which look at safety auditing implementation, monitoring and evaluation processes.

I would strongly recommend that all managers need to know something of the process of safety auditing so that they can play a more valuable part in this very important life-and cost-saving activity.

James Tye
Director General
British Safety Council

October 1992

PREFACE

This book is not an attempt to write a definitive work on safety auditing, the reason being that each industry will have to make auditing modifications in order to maximise the efficiency and effectiveness of this basic safety management tool. The purpose of this book is to discuss the safety management implications of safety auditing and to provide the reader with knowledge of some theoretical and practical aspects of safety performance management in order to encourage them to consider this aspect of their safety policy. Some larger organisations are already actively engaged in the safety auditing process but many smaller companies do nothing. It is hoped that this publication will provide a stimulus to those already involved in safety auditing and encouragement to those who are new to it.

Some safety auditing packages can look quite daunting to those unfamiliar with the subject and often this influences companies to carry on as before. Safety auditing does not have to be complex and some packages are quite simple and very easy to use. Care should be taken with those packages which award points or numbers for certain aspects of the safety audit process. This can sometimes give a false impression and boost management confidence unnecessarily. There is evidence of companies obtaining very high 'scores' at safety audit and a few weeks later suffering a terrible accident. Try to avoid this sort of numbering exercise wherever possible. Safety auditing should be as simple and straight forward as possible and should not look too complicated to the inexperienced eye. The best rule of thumb is to consider whether something has been done or carried out and whether it conforms to company policy. Here we stick as often as possible to either positives or negatives, but those who really like numbers can allocate a 0 or 1. It must be remembered that the purpose of the safety audit is to test whether clearly defined objectives have been met and this too could be recorded in a similar way.

This book has been written specifically for the manager or personnel officer who has a responsibility for health and safety and who:

- wishes to understand more about the safety auditing process;
- would like to carry out a safety audit of their workplace;
- would like to employ an external consultant to carry out a safety audit but would like to know more about it beforehand.

The methods considered in this book have evolved over a number of years in practice and the principles have been tried and tested in a number of companies worldwide. Developing countries have also been a part of the development programme. It is recognised that some modifications will always be required but the aim of the book is to provide a stimulus to those responsible for safety management. In this way, the safety audit can be seen as a tool for improving efficiency and effectiveness in the workplace.

R. A. Saunders, PhD   August 1992
Souchez
France

ACKNOWLEDGEMENTS

I am grateful to all those safety practitioners and safety managers in the private and public sectors who have made various comments, and for their valuable contributions to the contents of this book. In particular, I would like to thank Professor T. Wheeler at the Solent University and Dr D. Sheppard for reading the whole manuscript and for their valuable comments. I would also like to thank the staff at the Ministry of Communications of the Republic of Indonesia for their valuable support and particularly Ir. Manalom Hutahaean MSTr. and Brigadier General Darmawan Soedarsono for his help and support in testing some of the material in developing countries. I would like to thank the library staff at the Royal Society for the Prevention of Accidents, the British Safety Council and the Institution of Occupational Safety and Health for their research needed to undertake this work. I would like to thank my colleagues and friends for their encouragement and support whilst the manuscript was being written and to Ms Wisse for her kindness and moral support when it was most needed.

1 WHAT IS SAFETY AUDITING?

In this chapter we will look at the following:

- **Safety auditing – why?**
- **What is safety auditing?**
- **Introduction to safety auditing**
- **Auditing organisational safety policies**
- **Auditing safety procedures**
- **Auditing safety practices**
- **Auditing safety programmes**
- **The safety management process**
- **The safety mix**

SAFETY AUDITING – WHY?

There are many ways of attempting to measure an organisation's safety performance, some use financial criteria as the basis for the study whilst others use a more general approach which examines injury frequencies and so on. A safety audit is usually regarded as a measure of performance in terms of accident reduction/prevention, and examines safety policies, practices, procedures and programmes. Many organisations do not know whether their safety activities are good or bad, some not even knowing how much accidents cost their organisations in any one year. Others have an extremely dynamic and thorough understanding of safety management systems. The systematic investigation of all accidents in the workplace will usually be sufficient to indicate weaknesses in the management structure and to provide the basis for prioritising appropriate courses of action, allocating resources and identifying relevant counter-measures.

Safety performance might be regarded as a measure of the commitment of the workforce to the building of a safe and accident-free environment. In order to achieve this, every worker from the Chairman down must understand fully what is expected of them. Also, targets for both team and individual should reflect the strategic direction of the company, and employees must know when they are successful. Those involved in change must be given the opportunity to be involved in the management of the change processes. Attempts must be made to build upon successful performance and to encourage continuous effort to improve such performance. The aim of performance management, therefore, is to enable organisations to control activities connected with the safety of all involved in the achievement of the company business plans.

WHAT IS SAFETY AUDITING?

There is much talk about this part of the safety management process. Many students and managers often ask what exactly safety auditing is, and how and when it should be carried out. In discussion with clients and post-graduate safety management students it soon becomes apparent that most people have been evaluating, analysing and monitoring what they do, and naturally ask what the difference is between, say, safety programme evaluation and safety auditing. This obviously needs further discussion but safety auditing must start with the organisational mission in respect of health and safety. This might be a simple statement such as 'to reduce accidents and dangerous occurrences or to prevent them from happening'. Whatever the mission statement professes is what the audit will set out to test. Thus, it is necessary to examine objective statements at all levels and test whether these are adequate for the organisational mission to be realised. Some describe safety auditing as a measure of an organisation's safety performance. Such examinations may be carried out at all levels in the form of a review of organisational policy and planning, its procedures and systems, its practices and safety programmes. For the purposes of this book it is necessary to clarify some of the terminology and to examine some common terminology which are used in general safety management practice.

A safety audit will examine the 'whole organisation' from a health and safety perspective in order to test whether it is meeting its safety aims and objectives. It will examine hierarchies, safety planning processes, decision making, delegation, policy making and implementation, as well as all areas of safety programme planning, implementation, monitoring and evaluation. It will examine all socio-technical systems within the voluntary and statutory areas of safety management and will comment on their efficiency and effectiveness. A review, on the other hand, will normally examine any one part of the audit process such as a review of safety programme evaluation or a review of current safety procedures in a particular work area of the organisation.

Safety audits by their very nature involve a detailed examination of the whole organisation and can therefore take a considerable amount of time. In organisations employing 500 persons or more, safety audits should only be carried out every three years. If these are to be of any value to an organisation they should be conducted by persons experienced in safety management systems who are sufficiently divorced from the organisation to be capable of arriving at an objective and unbiased conclusion. In these larger organisations it would be appropriate to 'review' various safety activities annually, or as the accident situation dictates, concluding with a final 'audit'.

The smaller organisation, employing less than 500 persons, may present a simpler task and it may be possible to conduct safety audits more frequently. Each case would have to be judged separately taking into consideration the nature of the business the company was involved in and its accident record. Wherever a full safety audit is not possible then individual reviews of policy, procedure, practice and programmes would be expected.

INTRODUCTION TO SAFETY AUDITING

The systematic evaluation of policies, procedures, practices and pro-grammes is essential to the efficiency and effectiveness of any organisation. Many organisations carry out a form of *safety audit* or *review* once a year and this is seen as the mainstay of their safety policy whilst others evaluate only a part of their activities such as a specific programme. There are sever-al *safety auditing* packages available, some of which can cost a great deal of money. Like most off-the-shelf items, the package might not be entirely appropriate and some alterations might be necessary before it can be used. When carrying out an audit there are several methods recommended for assessing the effectiveness of the safety process. Some off-the-shelf pack-ages (e.g. CHASE) use a point-scoring system for each section of the audit-ing process. More traditional methods seek more quantifiable results such as detailed Economic Rates of Return (ERR) studies, Value For Money (VFM) exercises or Cost-Benefit Analysis (CBA). Whichever method is appropriate to organisational requirements the basic purpose of all these schemes is the regular examination of organisational activities. These should be examined quickly in order to provide the safety manager with simple answers to key questions in the following areas:

- policies;
- procedures;
- practices;
- programmes.

An organisation will need to know whether its policies are legally adequate; whether it administers and manages its responsibilities efficiently and effec-tively; whether current practices are providing adequate protection and whether accident reduction programmes are achieving organisational aims and objectives.

AUDITING ORGANISATIONAL SAFETY POLICIES

All organisations have responsibilities within various areas of legislation and it is important that the safety manager can quickly assess whether company policy is adequate in meeting its legal obligations. To do this it is necessary to obtain answers to the following basic questions:

Is the safety policy towards resourcing correct?

This aspect covers:

- Legal requirements – those rules and regulations covered by common and statute law, regulations and other legal requirements.
- Financial statement of objectives – organisational and individual. These should be agreed with those concerned.
- Budgetary provision – sufficient funds must be available to carry out agreed objectives from predetermined priorities.
- Staff – correct people should be deployed on the right tasks and training must be provided and monitored.
- Organisational structure – it is important that the hierarchical structure is capable of carrying out its aims and objectives efficiently, effectively and safely.

Is the safety policy adequate?

This aspect covers:

- Decision-making environment – safety decisions should be taken efficiently and effectively; they should be communicated to all affected personnel; policies, procedures, practice and programmes should be known to all concerned. This aspect would also consider objective setting and performance standard criteria.
- Safety committees – where appropriate these must consist of both senior management and interested worker representatives. Such committees must have sufficient authority to be effective otherwise they are just seen as 'talking shops'.
- Management involvement – senior management must be involved in all levels of the safety management decision-making process and must actively support the work of those specifically employed on accident prevention/reduction work.
- Trade union liaison – trade unions have a valuable and necessary contribution to safety management and should be actively encouraged to play their part efficiently and effectively.

- Liaison with other interested groups – effective accident reduction and/or prevention is the responsibility of everyone. Therefore the decision-making environment includes co-operation and discussion with all groups with an interest in the safety subject matter.

At this stage attempts should also be made to identify areas where difficulties might be experienced in terms of remedial action.

AUDITING SAFETY PROCEDURES

In this part of the safety auditing process, it is important to assess whether the administrative procedures currently in force are able to continue to implement organisational policy efficiently and effectively. Questions relating to this process concern the following issues:

- Administrative structure – this should be set up in such a way that it fully supports the health and safety policies, practice, procedures and programmes which are outlined and agreed as the company safety objectives. All too often professional staff have to conform to administrative procedures rather than the other way round. It is important to get this right.

- Technical and professional – procedures should be in place to meet the needs of any technical or professional aspects of tasks listed as objectives. These would include maintenance and repair procedures.

- Communication – all aspects of safety policy, procedures, practice and programmes should be disseminated in a sufficiently clear and concise form that they are understood by everyone. Steps should be taken at regular intervals to monitor all aspects of safety communication.

- Time management – safety requires time, thus it is important that working procedures allow for accident reduction and/or prevention measures to be taken at all levels within the organisation.

- Internal and public relations – it is important for both staff and people in the local community to be a part of the organisational safety activities, and procedures should be in place to cater for this.

- Recruitment – recruiting the right person for the right job is essential to the safety management system. Human failure is evident in over 80 per cent of accidents, therefore recruitment must involve appropriate procedures to ensure, as far as possible, that good staff are recruited and trained.

- Safety training – this is an essential part of the safety management system and procedures should be in place to examine current safety training procedures in terms of efficiency and effectiveness, and to ensure that they meet organisational goals.

- Supervision – there have been many examples given in recent cases where poor supervision has been identified as a major contributory factor in accidents. Supervisors should be subjected to good strong leadership and should have clearly defined areas of responsibility in terms of authority (i.e. the decisions that they can take without referral to a more senior manager) and accountability (for his or her actions).

- Discipline – the reasons for this, including the appropriate penalties which may form part of the disciplinary procedures within an organisation, should be communicated to, and understood by, everyone.

AUDITING SAFETY PRACTICES

This covers those safety practices which may be based upon historical precedence or upon professional ethics, codes of conduct and practice which directly involve organisational policy and procedure. Questions regarding this concern:

- Costing and valuation of accidents – it should be accepted practice that all accidents are costed and that an agreed methodology is employed to meet this aim. This is discussed further in Chapter 7.

- Accident investigations – many organisations do not adequately record their accidents or consider the practice of calculating accident costs to be part of their normal safety management system. Such information is essential for decision making and for the placing of objectives in order of priority.

- Data collection – the practice of systematically collecting accident data is an important part of the safety management process and the information obtained is crucial in calculating safety performance and for the purposes of comparison.

- Medical examinations – the practice of requiring certain workers to undertake certain detailed medical examinations is an important feature of the safety management process and is a requirement of the safety audit process.

- Welfare – this requires an enthusiastic commitment by management if it is to contribute effectively to health and safety practice in the workplace.

- Hazard and risk assessment – the practice of hazard and risk quantification and/or qualification should be encouraged at all levels, and safety systems should be in place to facilitate this. Included in this would be areas where accidents have been known to occur, where dangerous occurrences have taken place or where evidence exists as to the potential to cause injury and/or ill health. The practice of keeping up to date is an important feature of the audit.

- Accident analysis – companies should systematically analyse all accidents, however small, in order to identify the contributory factors which led up to the event. This will allow more efficient and effective countermeasures to be considered.

- Equipment inspections – the practice of regularly inspecting equipment must form an integral part of the safety management process. These examinations must be carried out in accordance with guidelines laid down by manufacturers or other competent personnel, and conducted by professionally qualified persons. Records must be maintained and an appropriate maintenance and repair procedure must be in place.

- HSE Codes of practice – these offer sound information and advice which has been developed through the experience of the HSE field staff, or gained from other professional bodies and organisations. They should be included in the safety management of an organisation where relevant and it is to be expected that they would form an integral part of the safety audit process.

- Professional Codes of practice – these are issued, from time to time, by bodies such as the British Medical Association, the Law Society, the Institution of Civil Engineers, the Institute of Occupational Safety and Health and so on. Members of such organisations are bound by additional rules, regulations and standards and as such, these rules, etc., should also be considered as part of the audit exercise.

AUDITING SAFETY PROGRAMMES

Here, the safety audit will examine past remedial strategies for efficiency and effectiveness using the four basic elements of the safety mix. These are:

- Enforcement – a football match without a referee is not a happy prospect. It is essential, therefore that laws, rules, regulations etc. are available, but they should be used in such a way as to complement other programmes being considered. Research tells us that the law should never be used in isolation and should always be used with at least one other feature within the safety mix (see Handbook of Safety Management, 1991). The law and how it is implemented is subject to audit.

- Engineering – engineering design, manufacture, equipment and machinery inspections and maintenance are all essential features of safety programmes and as such must form part of the safety audit. Like the law, research has shown that a programme of safety engineering used in isolation is not sufficient to reduce or prevent accidents from happening. A blend of socio-technical features used with other aspects of the safety mix have been found to be more effective in accident-reduction strategies.

- Environment – the working environment must be kept as safe and healthy as is reasonably practicable, and therefore must form an integral part of the safety audit process.

- Education – this term refers to safety education as a proactive means of developing skill, knowledge and safe behaviour. Safety training is a more reactive and readily available tool. Publicity programmes are included in this group, and therefore safety education, training and publicity programmes would be normally subjected to audit.

It must be remembered that all policies, programmes, procedures and practice form the basis of the audit if the *aim* of the safety audit or review is to improve organisational efficiency and effectiveness. Safety audit *objectives* would ultimately test an organisation's performance in terms of accident reduction and/or prevention and to keep risk at a minimum keeping the health and safety of the workforce at a maximum.

THE SAFETY MANAGEMENT PROCESS

Any manager charged with the responsibility of carrying out an organisation's health and safety policy must do so efficiently and effectively. To do this he must be aware of his duties, responsibilities and functions. To assist in this it is important to remember that there are seven basic tasks to consider. In being responsible for health and safety policy, procedure and practice in your particular workplace you should consider the following tasks:

Task 1 You will need to act as an accident investigator.

Task 2 You will need to plan policies, procedures and programmes designed to reduce accidents or prevent them from happening. These will be subject to safety auditing. Accident prevention includes all health matters.

Task 3 You will need to monitor regularly all health and safety policies, procedures, practices and programmes to ensure that they are adequate. Failure to do so will be highlighted at audit.

Task 4 You will be required to provide on-the-job instruction and training in the following:
- safe systems of working;
- workplace rules and regulations;
- employee responsibilities;
- indoctrination and education.

These issues will form part of the safety audit process.

Task 5 You will be required to provide and issue protective clothing in serviceable condition and/or equipment in accordance with the law and relevant regulations. You will also be required to maintain a safe environment in which you expect people to work. If you are in any doubt whatsoever about your duties and responsibilities you should know when and how to seek help. Such activities must form a part of the safety auditing process.

Task 6 You should get to know and understand your workforce. Do not expect an employee to do a job you know nothing about, you should be seen to lead by example. If hats are required on the shop floor then that rule applies to *all* employees including the personnel staff. Weaknesses here will be picked up at audit.

Task 7 You have a duty to ensure that health and safety matters are given a high priority within the decision-making process. It is important to establish the importance of health and safety at senior level and sufficient resources should be allocated to the tasks.

These tasks are summarised in Fig. 1.1 and are examined at audit.

The major areas of responsibility for safety managers cover the formulation, implementation and monitoring of appropriate action. Managers will have to organise other staff and as such are faced with the planning, organisation, motivation and control of staff. Since the introduction of the Health and Safety at Work Act (HASAWA) in 1974, organisations have had a statutory obligation to ensure the safety and well-being of their employees. This is a general piece of legislation covering broad safety principles, other forms of legislation such as the Factories Act 1961 or the Fire Precautions Act 1971 are more specific. Some organisations react more enthusiastically than others when implementing such legislation. Here, it is necessary to consider the implications of the HASAWA which placed broad statutory responsibilities on employers to ensure the health, safety and welfare of their employees at work.

In order to make decisions, information is required. There are a number of sources of information which are useful when planning, implementing, monitoring and evaluating health and safety policies, programmes and procedures. Useful information regarding what is happening in the workplace can be gleaned from:

– monitoring and evaluation;
– accident and dangerous occurrence investigation;
– personnel records;
– external sources.

Task	Specific activity
Accident Investigator	To carry out investigations into all accidents and dangerous occurrences in order to establish contributory factors
Advocate	Establish health and safety as a priority within the organisation and secure its recognition at board level. Secure sufficient resources.
Auditor	Carry out regular examinations of current health and safety policy, procedures, practice and programmes to ensure satisfaction.
Leader	Know and understand the workforce and lead by example. Motivate workers and develop schemes and plans to change attitudes and behaviour
Planner	Plan, implement, monitor and evaluate remedial measures designed to reduce or prevent accidents from happening.
Provider	Issue protective clothing and/or equipment. A knowledge of the legal requirements is necessary. Seek expert help where necessary.
Trainer	Provide on-the-job training, safe systems of working, indoctrination, workplace rules and regulations and establish employee responsibilities.

Figure 1.1 Safety manager's audited job tasks

Safety auditing is a crucial source of information. For the planning of remedial action, the manager will need to have access to various other types of information such as:

1.   The number of injury accidents which have occurred.
2.   The number of dangerous occurrences which have taken place.
3.   A sufficiently detailed breakdown of each incident that the cause and/or contributory factors can be identified.
4.   A sufficiently large database to allow the identification of trends.
5.   Supplementary information concerning accidents and dangerous occurrences from similar industries.
6.   Relevant national statistics where they are available.

It is important to realise that this type of information is a vital part of the decision-making process and increases efficiency and effectiveness in practice. Generally, personnel managers would need to collect detailed information concerning all dangerous occurrences in their respective organisations. This will include details of all injury accidents.

Previous studies have shown the poor quality of primary accident data collection procedures and that maximum use is not made of the information collected. Despite this, many of those charged with the responsibility for health and safety matters, continue to ignore or appreciate the value of a reliable primary data source. A strategic objective here, therefore, must include data collection and must ensure that these data are used for analysis purposes. This must form part of the decision-making process.

To do this within practical economic constraints, personnel managers should establish a system for collecting details of accidents and dangerous occurrences. It is known that many organisations spend considerable sums of money in order to meet a potentially dangerous situation. If an objective is to collect all accident information, then steps should be taken to collect this data in a sufficiently detailed form for potential problem areas to be identified. This will enable priorities to be established and ensure that resources are used effectively. At the same time, supplementary data should be systematically obtained in detailed, structured interviews with casualties so that lessons can be learnt. Wherever possible, these interviews should be conducted in total confidence and without fear of any consequent disciplinary action.

What other forms of information do we need? It has become established practice for the manager to be primarily involved only in certain specific areas of health and safety responsibility such as education, training and publicity strategies, whilst the engineer or scientist concentrates on engineering and some environmental issues. Such division should be avoided wherever possible. To enable organisational missions to be met at all, and in order to improve efficiency and effectiveness, both systems should work closely together since each of them plan and implement strategies using the same database for decision making. Both are equally dependent upon the quality of these data, thus there should be a shared and co-ordinated approach to remedying any problems of unreliability. It is from this premise that other forms of relevant information can be systematically gathered for programme planning and evaluation purposes. Information gathering, therefore, is of primary importance and should be given high priority within the objective setting sequence. Some aspects to consider are given in Fig. 1.2.

The majority of managers responsible for health and safety do not have direct access to any database for decision-making purposes. In any case, few have received training in its use. An unacceptably high number are unaware of its value and most have not considered whether their accident trends were a problem. Some managers collect accident data for establishing trends but tend to analyse individual cases rather than establish causes. Furthermore,

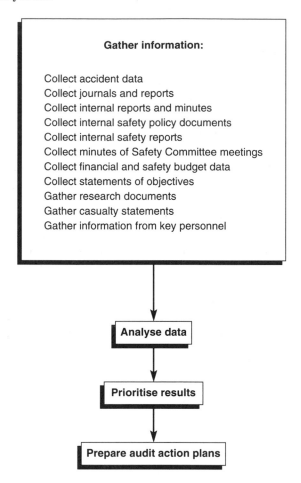

Figure 1.2 Sample safety audit preparation plan

safety managers do not regularly monitor dangerous occurrences nor are they familiar with statistical packages or research methodology. This is despite government departments, the HSE, the RoSPA and the BSC, regularly encouraging managers to evaluate their work. Few use basic research tools such as computers. The problem can be compounded because some organisations pursue staff recruitment policies that are not designed to address this problem. In too many cases, they fail to recruit numerate graduates for key management roles, or to provide for adequate staff development in the areas of information technology and statistics.

Evaluation can only occur within a framework of clearly defined aims and objectives and if undertaken by properly-trained staff. Many safety professionals

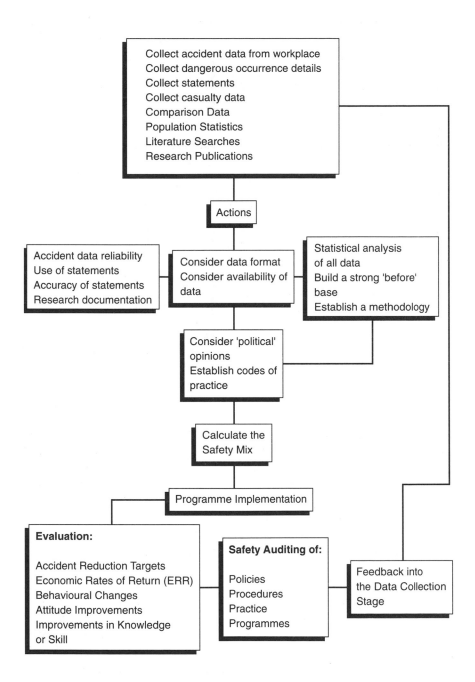

Figure 1.3 Safety management process

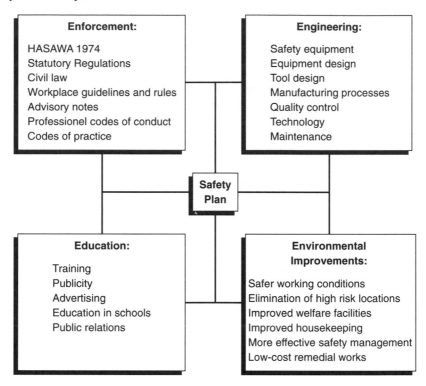

Figure 1.4 The safety mix

argue that safety activities cannot be evaluated, for example because you cannot quantify how many deaths or injuries have been avoided. This is symptomatic of this whole issue. Surely, if a particular plan were introduced to encourage the use of high visibility clothing in an unsupervised location, its results could be measured. If, after the campaign, more people are found to be wearing such material than before, then some degree of success may have been achieved, particularly if no such change occurred in the control group. Other areas of behaviour and attitude can be similarly evaluated. Management strategies can also be evaluated for efficiency and effectiveness. It is recognised that specific reductions in accidents directly attributable to an education, training and/or publicity programme are very difficult to establish in terms of a simple causal relationship. Nevertheless, the safety manager's contribution to the overall corporate approach in accident reduction is now acknowledged and its long-term value is becoming accepted. For example, a comparison of the number of accidents in relation to the vehicle population of this country would confirm this, particularly when compared with our EEC partners. The manager, therefore, can be an important and vital part of this corporate plan.

Whilst the safety manager might consider the planning, implementation, monitoring and evaluation of the safety mix programme within specified areas of operation, there are other specific tasks which he would also be responsible for. These are summarised in Fig 1.3.

For example, an assumption can be made from the above process, that all buildings and premises have already been occupied. Where a company policy is to relocate, then it is essential that policy plans are examined at the outset as part of the pre-audit exercise. This would examine all the physical aspects of the relocation and/or new business premises and would consider:

- building design;
- construction of the workplace and its environment;
- design, selection, purchase, siting and installation of plant and machinery;
- storing and use of hazardous materials and substances by employees;
- storing and use of hazardous materials and substances during the construction phase;
- construction work plan and examination of the critical path programme.

THE SAFETY MIX

The main aim of the manager responsible for health and safety is to co-ordinate the safety mix. This forms a main feature of the remedial process, and successful safety programmes are based upon a mix of some or all of the main ingredients shown in Fig. 1.4. It is very rare for one part of the mix to be successful in isolation. For example, the Department of Transport spent some £30 million attempting to persuade drivers to wear seatbelts. This was conducted via one 'E' of the safety mix, that involving education and publicity. It failed and a further 'E' was necessary in the form of enforcement. The two elements together now account for a 96 per cent wearing rate amongst car drivers. Similarly, speed limits, an enforcement requirement, are often ignored unless backed up by other measures within the safety mix framework. The task of the safety manager is to develop an appropriate blend within the safety mix framework at local level. This aspect is usually dealt with in the programme planning stages.

Care should always be taken not to rely solely upon the enforcement aspect of the mix as the only means of providing an effective accident prevention strategy.

It is necessary that all these tasks are carried out efficiently and effectively and within clearly defined cost restraints. It is far better for the safety manager to tell the accountants what is to be done rather than the other way round.

2 PERFORMANCE MANAGEMENT

In this chapter we will look at the following:

- **Performance management**
- **Uses of performance appraisal**
- **Safety performance standards**
- **Avoiding safety audit bias**
- **Rating scales**
- **Safety management orientated appraisals**
- **The implications of the safety appraisal process**
- **Evaluation interviews**
- **Summary of safety auditing and performance**
- **Safety audit management**
- **Financial aspects of safety performance**
- **Summary**

PERFORMANCE MANAGEMENT

It was mentioned in the previous chapter that safety auditing is the process of appraising safety performance. There are many facets to consider when examining performance management and the issues associated with it. Safety performance appraisal is a means by which organisations evaluate job performance in relation to health and safety within its accident reduction/prevention policies. If this is carried out correctly then employees, their supervisors and managers will obviously benefit from fewer accidents, dangerous occurrences and lower levels of risk. It is natural in any environment for employees to seek some form of feedback about their performance as this ultimately affects behaviour. This need is more obvious amongst new members of staff who are coming to terms not only with their new job but also with new colleagues and new surroundings. The more long term employees also need positive feedback on the good safe things that they do although they may not particularly like corrective feedback, this could come across as criticism and would require changes in behaviour.

Safety managers and supervisors need to evaluate performance in order to know and understand what corrective actions to take. Employee safety performance is usually compared implicitly or explicitly with set standards and the number and type of accidents and/or dangerous occurrences which have taken place since the last safety audit. This prompts the safety manager to reinforce desired outcomes and take corrective action for poor safety performance. An example of this might be placement decisions, promotions and dismissals rely on the appraisal of safety performance.

USES OF PERFORMANCE APPRAISAL

Improved performance provides the necessary feedback to allow an employee, supervisor and/or manager to intervene with appropriate actions to improve safety performance. Safety performance evaluations also help the decision makers to determine who should receive, for example, pay increases or safety performance bonuses, promotions, transfers, demotions or dismissals. However, poor safety performance may indicate a need to improve safety training whereas good safety performance may highlight some untapped potential which should be developed further.

Good or bad performance implies strengths and weaknesses in the personnel department's recruiting procedures in recruiting the right or wrong personnel for key jobs. However, poor performance might indicate errors in job analysis information, human resource planning or other parts of the personnel management information system. A reliance upon inaccurate information may lead to wrong appointments being made, inappropriate safety training being given or bad counselling decisions being reached. Also, poor performance may be the result of poor job design, therefore job appraisals should form an integral part of the safety auditing process. It is important to bear in mind that on occasions, performance is influenced by factors outside the working environment, such as family, health, financial or other personal matters. If these are picked up at safety performance appraisal time then there is a good chance that constructive help can be provided. Fig. 2.1 illustrates the elements which are considered to fall within an acceptable safety appraisal system. This approach should identify safety performance related standards, be able to measure those criteria and provide appropriate feedback to employees and managers throughout the organisation. If safety performance measures are not job related then subsequent evaluations could lead to inaccurate or biased results.

The Personnel Department usually develops performance appraisals, therefore the safety manager should liaise with this department when safety performance appraisals are to be undertaken as part of the safety audit process. There is no reason why this matter should not be delegated to that department provided that the safety manager supervises the process. The advantage of getting the personnel people to undertake this aspect of the safety audit is that it can produce a uniform approach to the exercise, provided that the safety criteria under examination are fully understood. Although the personnel department may develop a different approach to safety performance appraisal for

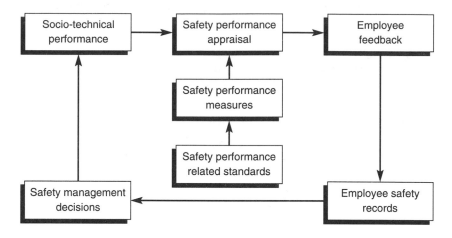

Figure 2.1   The key elements of safety performance appraisal

managers, professional staff and manual groups, it is important to have unifor-
mity within each group in order to ensure useful results.

Safety performance appraisal must create an accurate picture of an individual's job performance from a safety point of view. Such appraisal systems
should be task related and practical, have standards and should use reliable
measures. Job related means that the audit system evaluates critical behaviours that constitute job success. These behaviours are usually identified as
part of the safety job analysis process. If it is not task related then it is
invalid and this may discriminate between various groups of worker.

SAFETY PERFORMANCE STANDARDS

In order to evaluate safety performance, safety performance standards must be
set. These serve as benchmarks against which performance is measured. To be
effective, they should relate to the desired results of each task. These should
not be set arbitrarily and the knowledge of these jobs will be gained through
task analysis (see Chapter 4). An additional consideration is whether task
objectives are subjective or objective in nature. Subjective performance measures are those ratings which are not verifiable by others and rely heavily upon
opinion, whereas objective performance measures are those indications of task
performance that are verifiable by others. These are usually quantifiable. An
illustration of the accuracy of these measures is given in Fig. 2.2 below.

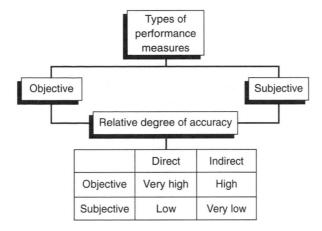

Figure 2.2 Types and accuracy of safety performance measures

This illustration shows that subjective safety measures are low in accuracy. When such measures are also indirect the accuracy becomes even lower. For example, the measurement of the safety department's telephone help-line and politeness might be done subjectively. This is because opinions are sought as to what constitutes 'helpfulness' and 'politeness'. Since this evaluation is subjective, accuracy is likely to be just as low if the safety auditor directly observes the safety department's help-line in operation. Accuracy is likely to be even lower when the safety auditor uses an indirect measure such as a written test on safety assistance and politeness. Wherever possible, the safety auditor should use objective tests and direct performance measures.

AVOIDING SAFETY AUDIT BIAS

Where performance measures are objective there can be little bias because these measures tend to be based on fact. It is where subjective issues are being safety audited that problems can arise. Such bias takes the form of inaccurate distortions of a particular measurement. It is normally created by safety auditors who fail to remain emotionally detached whilst they are evaluating a particular employee's performance. The following problems are the most common:

- Personal prejudice – where the safety auditor dislikes a group or class of people can influence the performance ratings received. Sometimes safety auditors are unaware of their biases and this is very difficult to overcome. This prejudice affects complete groups rather than individuals.

- Individuality effect – occurs when the safety auditor's personal opinion of the employee (the individual) influences the measurement of performance. An example would be whether the safety auditor personally likes or dislikes an employee and this affects the measurement of performance. This is most severe when the safety auditor and employee are either friends or strongly dislike one another.

- Cross-cultural differences – every safety auditor holds expectations about human behaviour and this is usually based on his own culture. Where safety auditors are expected to evaluate others from different cultural backgrounds, they are in danger of applying their cultural expectations to someone who has a completely different set of beliefs. In many developing countries, for example, the safety auditor has to be aware of the cultural differences, such as women in the Middle East who are expected to play a subservient role in public life. These women can easily receive a biased assessment when compared to more assertive women.

- Recent actions bias – when using subjective performance measures, the safety audit could be affected by the employee's most recent actions. Whether these actions are regarded as good or bad, they are more likely to be remembered by the auditor.

- Hard and soft biases – where leniency on the part of the safety auditor eases the performance standards. Some see all employees' performances as safe and assess them accordingly. Others are rather strict in their assessments, therefore it is important to set very clear performance standards.

- Central tendency problems – arise when safety auditors do not like to assess employees as effective or ineffective and an average performance is thus recorded. Such a distortion causes the auditor to avoid checking extremes and using classification such as excellent, very good, good, fair, poor and diabolical! Instead they tend to mark centrally. An example would be where an employee was required to undertake a periodic practical test. The practical examination or assessment form should be designed in such a way that the auditor is required to consider the extremes where necessary.

In order to reduce safety audit bias it is necessary to consider:

- training safety auditors to deal with subjective safety issues;
- using sound performance appraisal techniques;
- using objective safety issues wherever possible;
- designing clear performance appraisal record sheets;
- seeking a second opinion if in doubt;
- using possible extremes instead of creating central tendency problems;
- using accident and dangerous occurrence information.

Rating scales – job knowledge

Has a good knowledge of safety procedures and practice	5	4	3	2	1	0	Does not know anything about safety procedures and practice

Has a good working knowledge of health and safety law	5	4	3	2	1	0	Has no idea of the role of health and safety in the workplace

Rating scale of job performance

Reduced accidents in accordance with objective statement	5	4	3	2	1	0	Did not reduce accidents as identified in objective statement

Rating scales of individual characteristics

Initiative

Very high	High	Good	Average	Below average	Poor

Creativity

Very high	High	Good	Average	Below average	Poor

Figure 2.3 Rating scales

RATING SCALES

This is the most common system employed in organisations and provides a means of rating employees on each of a number of scales covering task requirements, performance and individual characteristics. Examples of some scales are given in Fig. 2.3. Some comments on these are as follows:

Paired comparisons

This provides a means of rating employees in comparison with each other. All employees of a similar level are compared with each other on a number of traits. This method avoids the problem of standards since the rating of employees is relative to each other. However, it can be a time-consuming exercise depending upon the number of employees being appraised.

Ranking

Staff are ranked in order of merit from best to worse. To use this system, as with the paired comparison method, criteria for measuring job performance will need to be generated.

Results centred

This approach relies upon some form of management objectives programme being in operation. At the beginning of a fixed time period, each subordinate agrees with his manager a number of performance targets to be met. The subordinate is then appraised by comparing his actual performance against the targets set.

Critical incident method

This system provides a means for managers to record critical behaviour on the part of their subordinates. Whenever an employee does something note-worthy, it is recorded on the appraisal form. These behaviours are normally classified by categories and then used to assess job performance. Critical behaviour will vary from job to job but examples could include such things as 'his weekly report was one week late' or 'he came up with an idea which could reduce accidents by 2 per cent'.

Forced choice

This system attempts to overcome the problem of the average rating by forcing the assessors to choose a statement which is most applicable to the particular individual being assessed. Groups of three or four statements are used such as:

- Administration: always keeps paperwork up to date; often leaves important letters to the last minute; filing is not very good.
- Toolroom: always leaves tools lying around.

There are other methods of appraisal and those mentioned are merely examples. One important point to consider when designing an appraisal scheme is to avoid it becoming administratively cumbersome and something regarded by subordinates as unimportant. The method of implementation and design

Competence	Grade A	Grade B	Grade C
Verbal communication skills	Able to communicate effectively with colleagues, clients, superiors and general workforce	Demonstrate fluency and confidence in dealing with professionals in related fields and to deal with press, radio and other forms of media on behalf of the organisation	To present reports at committee and formal meetings confidently, lucidly, persuasively and professionally and able to market ideas effectively
Written communication skills	Deal promptly and efficiently with routine correspondence and able to write clear and concise letters with minimum of supervision	Draft clear, consist and well argued reports with the guidance of senior colleagues	Ability to write clear, concise reports and letters without supervision
Personal effectiveness	Demonstrate a flexibility in approaches to meeting the need of the service	Ability to show effectively supervision and help with the general workforce	Ability to organise own workload and that of supporting staff and to show self-motivation in target completion. Show high standard of organisation and safety management
Motivation	Good	Very good	Excellent
Customer care	Demonstrate at all times tact, diplomacy and a commitment to the needs of the workforce and fellow professionals	Promotion of the safety service and policies, procedures, practice and programmes of the organisation at all times	Show an understanding and an application of the elements of good sound consulting skills
Organisational knowledge	Have a detailed knowledge of the local (and national) organisational structures and the key personnel within them	Understand fully the priorities and objectives of the safety department	Demonstrate a knowledge of effective safety systems and their management
Professional conduct	Adherence to professional codes of conduct and practice and to personal safety	Continuing to maintain the lower levels of competence and application	Continuing to maintain the lower levels of competence and application
Technical knowledge and skill	Ability to undertake and pass Part 1 of the NEBOSH or NEBROSH examinations	To undertake and pass Part 2 of the NEBOSH or NEBROSH examinations	To be a corporate member of the Institution of Occupational Safety and Health or International Institute for Risk & Safety Management
Numeracy	To demonstrate a good level of numeracy and ability to organise and present data effectively and accurately	Understand and be able to apply basic statistical tests to data sets	To be able to conduct rigorous statistical analysis on data sets and present findings accurately
Computer skills	Understand basic computer uses and demonstrate simple use of basic commands	Ability to load and run simple computer programmes necessary for the efficient and effective management of the safety office	Demonstrate good computer skills, demonstrate sound understanding of computer software and hardware

Figure 2.4 Performance appraisal of a trainee safety officer – level of competence

is important and it is essential that staff are aware of the reasons for the information being gathered and how it will help them. This will avoid next month's forms being filled in now!

An example of some performance standards required of a trainee safety practitioner is given in Fig. 2.4.

The appraisal interview

The most difficult part of performance appraisal is the interview. It is often rushed and takes place in an embarrassed atmosphere which then results in little benefit being derived from the exercise. The object of the appraisal interview must be to:

- let subordinates know where they stand;
- recognise good work;
- point out areas for improvement;
- indicate areas for further development;
- plan joint action for improvements;
- increase subordinates' motivation;
- improve management/staff relationships.

	Performance category
Excellent performance	You can rely on this manager to help all his customers in need
Good performance	You can rely on this manager to scale down the excited arguments which occur between keen executives over a half pint at lunch time
Fairly good performance	You can expect this manager to use discretion about whether to continue serving intoxicating drinks to young keen executives
Acceptable performance	You can expect the manager to stop staff serving those who are intoxicated and who have nobody to talk to
Fairly poor performance	You can expect this manager to take part in idle chit-chat with executives who are alone
Poor performance	You can expect this manager to ask everyone for their works identification before allowing anyone to be served
Very poor performance	You can expect this manager to encourage staff to hustle staff out of the premises whether or not drink or food is finished with little or no warning

Figure 2.5 A behavioural expectation scale for a company executive bar manager's customer relations

These objectives need to be borne in mind and fully understood by all parties to the interview if embarrassment is to be avoided. In order to improve the process, it is essential that the art of providing feedback is improved. Managers also need to appraise themselves and the art of good leadership lies in not giving people jobs that they are not prepared to do themselves and to show genuine interest in the subject matter and individual.

Behaviourally anchored rating scales

These are a group of evaluation methods that identify and evaluate relevant task-related behaviours. Since task-related behaviours are used, the validity of these methods is potentially greater than with other forced choice methods. The most popular of these is the behavioural expectation scales which use specific named behaviours as benchmarks to assist the safety auditor. This attempts to reduce the level of subjectivity and bias found in other approaches of performance measurement. An example of a behavioural expectation scale is given in Fig. 2.5.

SAFETY MANAGEMENT ORIENTATED APPRAISALS

Using the information gained from previous safety performance appraisals is an important part of the information gathering process, but there is merit in using the safety performance appraisal method in order to focus on future safety performance goals. There are four techniques normally used. These are:

- Self-appraisals – if you are attempting to improve self-development, then this method is a useful tool because defensive behaviour is less likely to take place and self-improvement is more likely. Whether you are using self-appraisal to look at the past or the future, the important thing to remember is the employee's involvement and commitment given to the exercise.

- Psychological appraisals – large organisations sometimes employ psychologists to assess future safety performance. Such a method employs in-depth interviews, psychological tests and discussions with supervisors in order to evaluate the employee's safety attitude, behaviour skill, knowledge, motivation, intellectual and emotional characteristics that may predict future safety performance.

- Objective setting – essential if employer and employee are to set realistic safety performance targets. Such targets must be measurable and if these conditions are met then employees are more likely to be motivated to achieve their goals since they took an active part in setting them. Also, since they are able to measure their safety performance they can adjust their behaviour to ensure that the agreed targets are met. It is important, however, that safety performance feedback is provided on a timely basis.

- Assessment centres – provide a method of assessing future safety potential, but they do not rely on one psychologist. In fact they are standardised employee safety appraisals that rely on several types of evaluation and auditor. Assessment centres are usually applied to those workers who have very responsible jobs. The process of assessment includes psychological tests, personal background histories, peer ratings, leaderless group discussions and simulated work exercises.

THE IMPLICATIONS OF THE SAFETY APPRAISAL PROCESS

Technique is not the only criterion necessary for good, effective safety performance. Irrespective of the method employed, it should be implemented among operators, supervisors and managers who have other priorities. Since these personnel may already know who their poor performers are, formal safety appraisals might not seem important or urgent. Successful safety appraisal almost certainly relies on managerial involvement and support and training might be required.

EVALUATION INTERVIEWS

These are review sessions that give employees feedback about their past performance or future safety potential. The safety auditor may provide feedback via several approaches:

- tell and sell;
- tell and listen;
- problem solving.

With the tell and sell approach, the employee's safety performance is reviewed and he is encouraged to perform better. It is the most common approach used with new employees. With the tell and listen approach, employees are permitted to explain reasons (or excuses) and defensive feelings about safety performance. It attempts to overcome some of these reactions encountered when counselling the employee on how to perform better. The problem solving approach identifies problems that could interfere with employee safety performance. Through training, coaching or counselling future safety goals can be set in order to remove deficiencies.

Emphasise positive aspects of employee performance

Inform each employee that the evaluation exercise is to improve performance, not to discipline

Carry out the safety performance review session in private with the minimum of interruptions

Review safety performance formally at least annually and more frequently for new employees or those who are performing poorly

Ensure that criticisms are specific and not general or vague

Focus criticisms on safety performance and not on personality or characteristics

Stay calm and never argue with the person being evaluated

Identify specific actions the employees can take to improve safety performance

Stress the evaluator's willingness to assist the employee's efforts and to improve safety performance

End the evaluation sessions by emphasising the positive aspects of the employee's safety performance

Figure 2.6 Some guidelines for effective safety performance evaluation interviews

Whatever approach is used to give employees feedback, there are some useful guidelines for effective safety performance evaluation interviews and these are given in Fig. 2.6.

SUMMARY OF SAFETY AUDITING AND PERFORMANCE

There are many facets to auditing safety performance. In order to begin an examination of your performance in practice, you should start by answering the following questions which are typical of those asked in management safety audits. The questions may be listed under the four main heads listed below in Fig. 2.7.

Consider your activity in each of these areas over, say, the last 12 months and answer these simple questions.

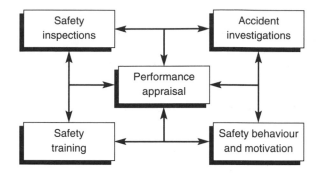

Figure 2.7 Main headings for safety performance appraisal

Inspections

- How many safety inspections have you made?
- How many unsafe conditions were found during these inspections?
- How many of these unsafe conditions were corrected as a result?
- How many unsafe behaviours were observed?
- How many of these behaviours were corrected?
- How many unsafe conditions were reported to you?
- How many of these reports did you respond to?

Your answer to these questions will be some indication of your personal involvement in hazard spotting and problem correcting in the area for which you are responsible. You should be taking the lead in the supervision of safety and ensuring that your staff comply with the safety rules.

Inspections should be carried out frequently but how often really depends upon the magnitude of risk involved in the work. Safety representatives are entitled to carry out their own inspections and you should be prepared to respond to their reports appropriately.

Accident investigations

- How many accident investigations did you carry out (in relation to the number of accidents)?
- Were these investigations made promptly or were they made subsequent to an injury benefit claim or safety officer's enquiry?
- How many times did you discover the true cause of the accident? For example, if you ascribed the accident to operator error what factors such as training, supervision, etc. led to this error?
- How many causes could be attributed to failure in the management system?
- How many causes were remedied successfully?

Before choosing to investigate an accident, a number of factors are generally taken into account such as the frequency of the occurrence, the severity of the injury, or just the unusual nature of the incident.

In fact there is a duty on all employers under the Social Security (Claims and Payments) Regulations 1979 to take reasonable steps to investigate the circumstances of all accidents notified to them. This investigation should be the starting point in seeking the root causes leading up to the accident itself. If you have relatively few accidents there is no reason why you should not make a serious investigation of them all.

SAFETY TRAINING

- How many new employees were trained in safety matters, for example, fire procedures, first-aid facilities, accident or hazard reporting, safety rules, etc?
- How many existing employees have had refresher training in safety matters?
- How many safety briefing sessions have you given or arranged?
- How many of your staff have attended safety training courses?
- What improvements have resulted from this training?

If you found it difficult to answer these questions, it may be because you have no tangible record of what has been done over the last year. It is important to keep a record of all training done, however small, for without this it will be very difficult to demonstrate compliance with the HASAWA.

SAFETY MOTIVATION

All managers will have a mission to make maximum effective and efficient use of resources with the maximum degree of health and safety. The concept of efficiency is seen as 'input centred' because it is concerned with ensuring that activities are carried out in a prescribed and appropriate fashion. Effectiveness, on the other hand, is seen as 'output centred' because it is concerned with the extent to which useful achievements are accomplished. An example list is provided in Fig. 2.8 and you are asked to consider whether the points listed are 'input' or 'output' centred.

The rationale underlying the need for safety managers or supervisors, who have a responsibility for health and safety, to have a clear understanding of their objectives rests on a distinction that can be made between the

Topic	Input centred	Output centred
Internal safety audit in your opinion?		
Ensuring employees arrive/leave work on time?		
Expecting tidy work areas?		
Safety appraisal interviews?		
Task analysis exercises?		
Objective setting?		
Safety budgetary control?		
Accident reduction targets?		
Expecting safety staff to be busy and hard working?		
Expecting safety equipment/clothing to be used/worn?		
Official safety procedures?		
Safety committee procedures?		
Identification of safety training needs?		
Safety management development courses?		
Safety publicity exercises?		
Internal safety campaigns (in your experience)?		

Figure 2.8 Input/output exercises

concepts of efficiency and effectiveness. Effective managers:

- *Do right things*   rather than *do things right*
- *Produce safe alternatives* rather than *accept tidy solutions*
- *Optimise resource use*   rather than *safeguard some resources*
- *Get results*   rather than *discharge duties*

Whilst a manager might be regarded as efficient he might not be very effective when output results are measured. It is therefore necessary to set objectives so that areas of efficiency and effectiveness might be appraised within the framework of health and safety requirements both voluntary and statutory. The consequence of not setting realistic objectives would be that an absence of goals would lead to input-centred behaviour and role ambiguity and conflict being experienced.

Safety auditing, therefore, must examine objectives and the performance of personnel within its policies, programmes, procedures and practices in terms of efficiency and effectiveness.

SAFETY AUDIT MANAGEMENT

It must be remembered that the *aim* of the safety audit or performance review is to improve organisational efficiency and effectiveness whilst its *objectives* are to reduce or prevent accidents from happening, to keep risk at

Performance factors	Far exceeds job requirements	Exceeds job requirements	Meets job requirements	Needs some improvement	Does not meet requirements
Quality	Leaps tall obstructions with a single safe leap	Requires a running start	Can only jump obstructions without barbed wire	Crashes into obstructions but tries hard	Cannot recognise obstructions let alone jump
Timeliness	Is faster than a speeding train	Is as fast as a speeding train	Not as fast as a speeding train	Tries to keep up with British Rail	Runs into himself
Initiative	Is stronger than a bulldozer	Stronger than a bull elephant	Stronger than a bull	Smells like a bull	Is full of bull
Adaptability	Walks on water	Walks on water but only in desperate situations	Washes with water	Drinks water	Passes water in dangerous situations
Communication	Talks to God	Talks with saints	Talks to anyone	Talks to himself	Argues with himself but loses those arguments

Figure 2.9 A light-hearted performance appraisal criteria for 'Supersafe' the Personnel Manager

a minimum and the health and safety of the workforce at a maximum. For safety auditing to be effective it should be carried out by a person who has:

- some independence from all departments being audited;
- sufficient seniority and authority to carry out the task without hindrance;
- knowledge of health and safety statutory and voluntary requirements;
- ability to articulate findings to senior managers.

It must also be remembered that individuals and their roles within the organisation must form an integral part of the safety auditing process. A light-hearted individual appraisal form is given in Fig. 2.9.

The framework for the individual manager's health and safety objective setting is provided by a document known as the safety manager's guide. This differs from a job description in that it concentrates on the results the person is expected to achieve (outputs) rather than the activities he will undertake (inputs). The guide must identify the main purpose of the job's existence in the organisation, the key results that must be achieved to fulfil the main purpose and the means by which one can judge whether these key results can be assessed.

At the end of an agreed timescale, the performance of individuals is reviewed against the results and targets contained in the safety manager's guide. In this way, the basis of the audit is performance rather than personality. Also, some relatively objective data is available in conducting the review. The safety manager's guide is, therefore, a document used to analyse the results which are expected of everyone in sufficiently precise terms to be useful for safety auditing purposes. The safety manager's guide will consist of:

- main purpose of the job;
- key result areas;
- key tasks;
- standards of performance;
- control data;
- improvements.

Main purpose of job

This contains a brief statement of the reasons for the existence of the job within the organisation. It should not be a description or summary of the activities involved but should indicate the results or benefits that the job role contributes to the organisation. As a general rule, the main purpose statement should only include one verb since to have more than one implies that the statement does not describe a single main purpose of the job.

In terms of health and safety a typical safety manager might state that the main purpose of his job is:

to prevent accidents from happening

Key result areas

In every job, there will be a limited number of distinct areas in which effective performance makes a significant impact on the successful achievement of the job purpose. These are referred to as the key result areas.

A key result area is an area of work which is critical to the continued success of the main purpose of the job. Alternatively it is an area where, if the job were not done well, there should be a significant deterioration in the quality or quantity of work and results. One key result area for the safety manager might be:

the safety training of all staff

Key tasks

Within each key result area one or more key tasks can be identified and defined in terms of specific actions and results that those actions are purported to achieve. For example:

- to control..........;
- to plan.............;
- to determine........;
- to develop..........and so on.

A key task might be:

to ensure that the safety training budget is not overspent

Standards of performance

Identifying the effect of an activity assists in defining the required level of performance, and leads the way to transforming generalised statements of desirable objectives into specific targets against which success can be measured. Standards of performance are statements of the conditions that will exist when the results have been achieved satisfactorily. They should not be set at ideal levels, but rather at realistic ones. They must be feasible and be standards against which it is appropriate to judge the employee's performance. One key task may have several performance standards such as:

- quantity – how many, how often?
- quality – how good, how safe?
- time – by when?
- cost – at what cost?

Wherever possible, performance standards should be measurable or objective terms such as:

Some 'good' examples of performance standards:

- maintain all safety equipment;
- produce safety literature and advice free from errors in grammar, spelling and punctuation;
- avoid overflowing rubbish bins;
- keep safety equipment service time to not more than 10 per cent of working hours;
- reduce safety maintenance to not more than 15 per cent of working hours from 20 per cent last year.

Some 'bad' examples of performance standards are:

- maintain safety equipment when requested;
- maintain a high standard of English in safety literature;
- empty waste bins twice a day;
- keep safety equipment serviceable at all times;
- reduce safety maintenance time.

When writing good objectives, an attempt should be made to include as many of the four elements of quality, quantity, cost and time as possible. It is unlikely that a meaningful objective would contain less than two of these elements.

Standards of performance in relation to the safety training (key result area) and financial control of the budget (key task) might be:

- information for the preparation of the annual health and safety budget obtained by (DATE) of each year;
- annual estimates prepared by (DATE) each year;
- financial control statements examined at the end of each accounting period in order to ensure that:

  – the estimates are revised immediately a new trend becomes apparent;
  – the difference between the actual expenditure and estimated expenditure does not exceed x per cent of the estimate or £y (whichever you have decided).

Control data

The next stage in the process is to identify the information that will be used to assess whether the performance standards are being met and also to identify the documentary sources that will be used to provide this information. This may be obtained from existing control documents, routine statistical information (such as accident data), reports and so on. Control data must be:

- accurate;
- available when needed;
- available in the right form.

The questions to be asked at this stage are:

- do you have sufficient feedback on progress in the system?
- do you have continuous feedback concerning resource use?
- have you ensured continuous comparison of performance?

In some instances no adequate control data may be available. In these circumstances one must estimate the time it will take to obtain this information then readjust the criteria accordingly.

Improvement of results

These are the recommendations for changes which are to be made as a result of the safety audit in order that higher standards of performance in key areas are achieved. The types of action that these recommendations are intended to set in motion are:

- improvements in efficiency (resource use);
- changes in procedure or practice;
- further safety training, education or publicity;
- investigation into problem or high risk areas;
- action to improve or change organisational structure.

The questions a safety auditor might ask at this stage are:

1. What facets of the job cause the most problems and accidents and what needs to be done, by whom, to bring about a worthwhile and lasting change?
2. What is currently preventing higher standards being achieved?
3. Would any of the work be simplified or made safer if procedures or practice were altered or another section or department did something different?

4. What changes, if made, would make tasks easier or less risky to perform?

5. Where could appreciable savings be made and how?

FINANCIAL ASPECTS OF SAFETY PERFORMANCE

Most safety programmes, including the safety audit, operate from a budget allocated for carrying out company safety policy. A system of budgetary control establishes various budgets which set out in financial terms the responsibilities of management in relation to the requirements of the overall policy of the organisation. There should be a regular comparison of actual results with budget forecasts both to try to ensure (through action by the safety manager) that the objectives of safety policy are met and to form a basis for any revision of such policy.

The crux of the budgetary process is that financial limits are allocated to component parts of the organisation. Thus, the safety manager plans activities in line with company policy and within the financial limits. It is important to try to obtain departmental management agreement to this financial limit as it is the manager who is going to be held responsible for keeping within it. Experience shows that a safety manager is more willing to accept responsibility for performance against his budget if he has been allowed to participate in the determination of the size of that budget rather than it being imposed upon him.

Primary responsibility for the administration of the budgetary process is normally delegated by senior management to a budget accountant who has the task of co-ordinating the preparation of both the budgets and their reports. These reports may be presented to a special budget committee (particularly in larger organisations) which is composed of the managers in charge of the major functional areas of the organisation. Membership of this committee would normally be extended to the safety manager.

Preparation of budgets

The following steps are typical of those taken in the preparation of the individual budgets and master budget for a commercial organisation. A safety manager responsible for safety at various locations, such as a group safety manager, would be responsible for his own master budget. This would then form a part of the overall organisational master budget:

1. A statement of overall safety objectives is prepared on which the individual budgets are to be based.

2. Forecasts are made regarding the general economic conditions and the conditions likely to be prevailing in the industry. Here accident data plays a key role.

3. A safety budget can then be prepared based upon the forecasts and will highlight key task areas for action.

4. A production budget is prepared in conjunction with the above and will require consideration of all materials and other resources required to carry out those key task areas presented in 3 above.

5. The administrative cost budget is prepared for each area of activity.

6. A capital expenditure budget is compiled covering anticipated changes in legislation, or where specialised equipment or modifications are involved.

The preparation of the budget is shown diagrammatically in Fig. 2.10.

Budgetary control during the year

Once the year has started, the control aspect of the budgetary process consists of comparing the actual results to budgeted figures. The chief value of the budgets, as a control mechanism in this connection, will be achieved through the effective use of regular reports. These reports, co-ordinated by the budget accountant, will show the variances (normally expressed as percentages) between the actual and budgeted figures. Significant variances should be highlighted. Variance control charts can be employed to present control limits to safety staff and these charts will also show up the significant variances as actual results are recorded.

The reports should also make it clear which budget variances were controllable by those responsible and which were not. Detailed explanations of the cause of the variances can be based upon the type of variance analysis between standard unit costs and actual unit costs. It must be remembered that budgetary control needs action. The chief value of the reports lies in effective use of reports. Budgetary control thus employs the concept of what may be described as 'responsibility accounting'. The accounting system in operation must provide information in line with the budgetary system (appropriate cost centres). In this context these issues will form the basis of the safety audit.

The most important aspect of financial performance at audit is in regard to the policy and procedure for costing an accident. There are many methodologies employed by companies but the one put forward here is that tested by the HSE and is discussed in more detail in Chapter 7.

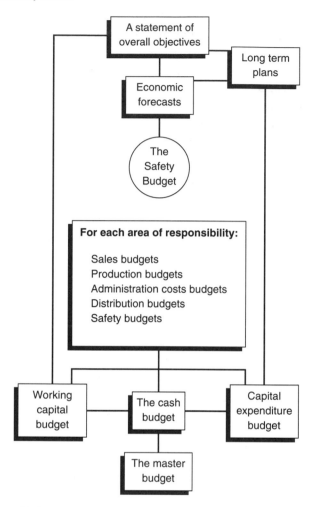

Figure 2.10 Budgetary preparation

SUMMARY

It is essential that performance management considerations form part of the company safety policy documentation. Without setting out clear and precise safety standards within the objective setting phase of the decision-making process it will be almost impossible to carry out a meaningful safety audit. Part of this process includes having a good understanding of the roles of the various members of staff who will be a part of the safety audit and task analysis data should also form part of the exercise. Attempts should be made to identify clearly the performance standards within each area of the study so that these may be more easily appraised.

All too often, safety audit programmes merely commence with a list of questions pertaining to various working situations. Many of the enquiries usually rely upon the opinion of the safety auditor for their answer, and sometimes this particular view can lack a detailed understanding of the roles and tasks of each individual under scrutiny. In such cases independent safety auditors should seek the help of the personnel department before proceeding any further.

Performance monitoring should take place through a variety of means such as strategy review, safety audit, operating plan reports and special investigations and through regular informal meetings and discussions. Without specific markers, there is less formality in the monitoring of strategic progress and more flexibility in the items which receive most attention.

Research now tells us that the more difficult goals can lead to better performance than the easier goals. This is not just a matter of getting more performance out of the lazy and incompetent. It also appears possible for managers to respond to the challenge in stretching goals and produce surprising results. However, overly difficult goals must not be set as this will deter people and can actually reduce performance. Safety management goals must be accepted rather than rejected as an unachievable proposition. Targets must be set which are achievable.

Given that strategic performance is important to an organisation the question then arises of how should the performance be rewarded. What alternative forms of safety performance reward can contribute to the company's long term competitive position. Bonuses based on accident savings can be considered as an acceptable form of reward but can be reduced if targets are not met.

Performance management is the process of improving business performance by directing attention towards the key areas of activity. It is a system which encourages greater focus on corporate safety objectives and resultant business and personal objectives. This is achieved by clarifying, at each level of management, what the key safety objectives are, and rewarding staff accordingly on achieving them.

The core principle of safety performance management is that it should link strategic operational plans to departments, units and individual goals. It is based upon agreed targets and the agreed focus on key results, placing emphasis on self-measurement and self-development. Safety performance problems are identified and the difference between targets and achievement must be recorded. The process requires the successful reinforcement of safety performance and must become the vehicle for continuous performance improvement.

Safety performance measures are ways of measuring the achievement of goals. The selection of appropriate methods of measurement will help the post-holder and appraiser to assess safety effectiveness. There are seven points to consider:

- Accountability or safety goals;
- Safety performance measures;
- Safety performance standards;
- Safety targets;
- Action plans;
- Progress reviews;
- Safety performance reviews.

The accountability or safety goals should clearly define the areas of responsibility of the post-holder. These are discussed in more detail in Chapter 4.

3 ROLE ANALYSIS

In this chapter we will look at the following:

- **Introduction**
- **Management tasks**
- **Role negotiation**
- **Responsibility charting**
- **Force field analysis**
- **Gestalt orientations**
- **Group behaviour**
- **Individual role analysis**
- **Job titles**
- **Job purpose**
- **Responsibilities**

WHAT IS YOUR MANAGEMENT TASK?

Role analysis is a process that collects information about jobs. This is done via surveys, observation and discussions amongst employees. From this information, role and position descriptions can be produced. This information can also provide the basis for the formulation of safety task standards which are required for safety performance appraisals and compensation details. However, role analysis techniques (RAT) are also designed to clarify role expectations and obligations of team members in order to improve team safety effectiveness. Within organisations, individuals fill different specialised roles in which they manifest certain behaviours. This division of labour and function facilitates organisational safety performance. Often the role incumbent may not have a clear idea of the behaviours expected of them by others, and what others can do for them to assist the incumbent in fulfilling the role may not be fully understood. There have been techniques developed which help to clarify the roles of management and roles where ambiguity or confusion exists. The intervention is predicted on the basis that consensual determination of role requirements for team members consists of a combined building of all the requirements by all concerned. This leads to a more mutually satisfactory and productive behaviour.

In a structured number of steps role incumbents, in conjunction with team members, define and delineate role requirements. The role is defined as the *focal role*. In a new organisation, it may be desirable to conduct a role analysis for each of the major roles. The steps involved are:

Step 1   An analysis of the focal role which is initiated by the focal role individual. The role, its place within the organisation, the rationale for its very existence and its place in achieving the overall organisational safety goals can be examined along with the specific safety duties of the post. The specific duties and behaviours should be written on a chalkboard and should be discussed by the entire team. Safety behaviours are added and deleted until the group and the role incumbent are satisfied that they have defined the safety role completely.

Step 2   An examination of the focal role incumbent's expectations of others. The incumbent lists their expectations of the other roles in the group that most affect the incumbent's own role performance and these expectations are openly discussed, modified, added to and finally agreed upon by the whole team.

Step 3   A discussion of others' expectations and desired behaviours of the focal role. The members of the team should describe what they want from and expect from the incumbent in the focal role. These expectations of others are openly discussed, modified, added to and agreed upon by the entire team including the focal role holder.

Step 4   On conclusion of Steps 1 to 3, the focal role holder assumes the responsibility for making a written summary of the role as it has been defined. This is sometimes referred to as the *role profile* and is derived from the results of the discussion groups. The role profile should contain the following information:
- a set of activities classified in relation to the prescribed and discretionary elements of the role;
- the obligations of the role to each role in its set; and
- the expectations of the role from others in the set.

This provides a comprehensive understanding of each individual's role space. The written role profile is briefly reviewed at the following meeting before another focal role is analysed. The accepted role profile constitutes the role activities for the focal role person. This intervention can be non-threatening with good results. All too often, the mutual demands, expectations and obligations of inter-dependent team members may never have been openly examined. Each role incumbent may ask why someone is not doing what they are supposed to do, while in reality all the incumbents are performing as they think that they are supposed to. Collaborative role analysis and definitions by the entire team will not only clarify what is expected of each but will ensure commitment to the role once that it has been clarified.

This process can be reduced considerably if there is already a high level of understanding of the current activities of the various role incumbents.

ROLE NEGOTIATION

When the causes of team ineffectiveness are based on the individual's behaviours which they are unwilling to change because it would mean loss of power or influence to the individual concerned, role negotiation can often be used. Role negotiation intervenes directly in the relationships of power, authority and influence within the team. The change effort is directed at the work relationships among members. It avoids probing into members' personal feelings about one another. The technique is basically an imposed structure for controlled negotiations between parties in which each party agrees in writing to change certain behaviours in return for changes in behaviour by the other. The behaviours relate specifically to the job.

The role negotiation technique usually takes at least one day to conduct so time must be set aside in order to carry it out properly. The process is summarised as follows:

Step 1 Contract setting – here the manager sets the climate and establishes the ground rules. It is important to look at safety behaviours rather than feelings about people. Be specific in stating what you want others to do more of or do better at, stop doing or carrying on unchanged. All expectations and demands must be written. Nobody should agree to change any particular behaviour unless there is complete agreement in which the other must also agree to any change. The exercise consists of individuals negotiating with each other in order to arrive at a written contract of what behaviours each must change.

Step 2 Issue diagnosis – here individuals consider how their own effectiveness can be improved if others change their work behaviours. Each person completes an issue diagnosis form for every other person in the team. On this form, the individual states what they would like the others to do more or less of, or maintain unchanged. These messages are then exchanged among all members and the messages received by each person are then made public amongst the group.

Step 3 Negotiation period – is where two individuals discuss the most important behaviour changes they want from the other and the changes they are willing to make themselves. Agreement is required, as each person must give something in order to receive something. Often this step is illustrated by two people negotiating in front of the entire group. The group then breaks into smaller negotiating pairs. This process consists of parties making contingent offers to one another such as 'if you do X, I will do Y.' The negotiation finishes when all parties are satisfied that they will receive a reasonable return for whatever they are agreed to provide. All agreements must be written and it is best to provide follow-up meetings in order to determine whether contracts have been honoured and to assess the effects of the agreement's effectiveness.

RESPONSIBILITY CHARTING

In working groups decisions are made, tasks are assigned and individuals and small teams accomplish tasks. This can be simply described on paper but in reality a decision to have someone do something is somewhat more complex than it appears because there are multiple factors involved in even the simplest of task assignments. There is the person who does the work, one or more people who may approve or veto the work as well as the persons who may contribute in some way to the work while not being responsible for it. The point is, who is to do what, with what sort of involvement with others?

A technique referred to as responsibility charting helps to clarify who is actually responsible for what in various decisions and actions. It is simple, relevant and very effective for improving group functioning. The first step is to construct a grid which contains the types of decisions and classes of actions that need to be taken in the total area of work under discussion. These are listed along the left-hand side of the grid and the actors who may play some part in the decision-making process are listed along the top (see Fig. 3.1). This process is one of assigning behaviour to each of the actors opposite each of the main issues. There are five classes of behaviour and these are:

1. Responsibility (R) – the responsibility to initiate action to ensure that the decision is carried out. For example, it would be the safety manager's responsibility to initiate the company safety policy.

2. Appeal needed (A) – the particular item must be reviewed by the particular role occupant and this person has the choice of either accepting or rejecting it.

3. Support (S) – providing logistical support and resourcing for the particular item.

4. Information (I) – must be informed and by inference cannot be influenced.

5. Non-behaviour (N) – is the non-involvement of a person with the decision.

Responsibility charting is normally carried out in the work group context. Each decision or action is discussed and responsibility is assigned. Researchers have provided guidelines for making the technique more effective. These are:

Types of decision to be taken	Classes of actions			
	R	A	S	I
Company safety documentation				
Accident investigation				
Accident costing				
Safety auditing				
Task analysis				
Role analysis				
Codes of practice				
Administrative procedures				
Safety training				
Safety publicity				
Budgetary control				
Budget expenditure				
Safety programme planning				
Accident costing				
Formulating company safety policy				
Staff recruitment				
Safety communication				
Operations management				
Accident analysis				
Counselling				
Welfare				
Dangerous occurrence reporting				
Safety maintenance				
Safety equipment and clothing				

Figure 3.1 A responsibility chart

- Assign responsibility to only one person. That person initiates and is then responsible and accountable for the actions.

- Avoid having too many people with an approval-veto function on an item. This will slow down task accomplishment or will negate it completely.

- If one person has approval-veto involvement on most decisions then that person could become a bottleneck for getting things done.

- The support function is critical. A person with a support role has to expend resources or produce something that is subsequently used by the person responsible for the action. This support role and its specific demands must be clarified and clearly allocated.

- The assignment function to persons become difficult at times. For example, a person may require (A) on an item but not really need it. Likewise a person may not like (S) responsibility on an item but should have it; or two people might want (R) but only one may have it.

A responsibility charting exercise can readily identify who is to do what on new decisions as well as assist in pinpointing why old safety decisions are not being accomplished as required. Responsibility charting is a reliable intervention to incorporate as a means of improving the safety task performance of a particular work group.

FORCE FIELD ANALYSIS

This is one of the most tried and tested forms of organisation and development analysis. It is a device for understanding a problem situation and taking corrective action. The technique is based on several assumptions:

- the present state of things;
- the desired state of affairs;
- stabilising the situation.

The current condition is a quasi-stationary equilibrium representing a resultant in a field of opposing forces. The desired condition can only be achieved by dislodging the current equilibrium, moving it to the desired state and stabilising the equilibrium at that point. In order to move the equilibrium from the current to the desired state the field of forces must be altered by adding driving forces. It is basically a vector analysis. Force field analysis involves the following steps:

Step 1   Decide on the problem situation that you are interested in improving and carefully describe the current condition in detail.

Step 2   Carefully and systematically describe the desired condition.

Step 3   Identify the forces and factors operating in the current force field and identify those which are pulling in the direction you want to be moving in.

Step 4   Identify the restraining forces which push away from the desired condition. Identification and specification of the force field should be thorough and exhaustive so that the picture of why things are as they are becomes clearer.

Step 5   Examine the forces and identify which ones are the strongest and which are more susceptible to influence. Identify which forces are under your control and which are not. Important individual forces could themselves be subjected to a force field analysis in order to understand them better.

Step 6   Strategies for moving the equilibrium from the present position to that which is desired are the following:
– add more driving forces;
– remove restraining forces;
– do both of the above.

There are those that advise not simply to add new driving forces because that could increase resistance and tension to the situation. Therefore, in this step, one selects several key, adaptable restraining forces and develops action plans to remove them from the field forces. As the restraining forces are removed, the equilibrium shifts towards the desired condition. New driving forces may also be proposed and action plans developed to implement them.

Step 7   Implement the action plan and this should cause the desired condition to be obtained.

Step 8   Describe what actions must be undertaken to stabilise the equilibrium at the desired condition and to implement those actions.

An example of force field analysis is given in Fig. 3.2. Assume that the Blogg Manufacturing Company are concerned at the number of reported dangerous occurrences and want to correct the situation. They would generate the following analysis.

Outlining the field forces in this way allows management to understand the many features of the problem. They would then decide which of the restraining forces should be removed and allow them to develop action plans capable of initiating change. This technique is a good way of diagnosing change situations. Although it is referred to sometimes as a team or group intervention it is also a useful tool for individuals. Team analysis usually yields a comprehensive understanding of what is happening to cause problems and what action is required to rectify them.

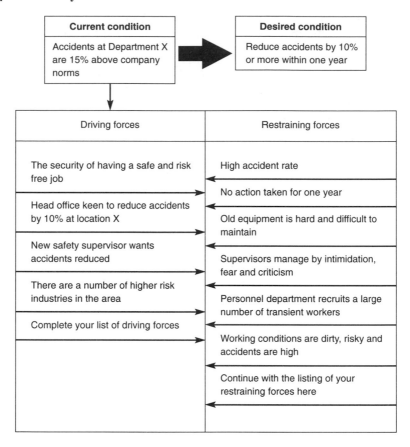

Figure 3.2   Force field analysis of an imaginary accident problem in Department X

GESTALT ORIENTATIONS

This approach concentrates on the individual rather than the group. It rests on a form of psychotherapy and is based on the belief that people function as a whole, complete organism. Each person possesses positive and negative characteristics that must be accepted and allowed expression. People tend to experience problems when they get fragmented, when they do not accept their complete selves and when they are trying to live up to the demands of others rather than themselves. The goals of Gestalt therapy are:

- awareness;
- integration;
- maturation;
- authenticity;

- self-regulation;
- changes to behaviour.

People must come to terms with themselves, they must accept responsibility, must be realistic and not block awareness, authenticity or any of the dysfunctional behaviours. The aim is to make the individual stronger, more authentic and more in touch with his or her own feelings. This may help to build a better group but it should not be the main outcome. The objective is to help the individual to recognise, develop and experience their own potency and ability to cope with the work situation whatever its current condition. Furthermore, individual workers should be encouraged to discover for themselves their own unique environment and capacity to influence and mould it in such a way that they can achieve more of what they desire.

In order to do this, people must be permitted to express exactly how they feel about something whether it is positive or negative. They must learn to get involved with issues that affect them, understand their relations and responsibilities towards others and work through problems until they are solved rather than just expressing negative feelings. People must learn also to accept their strengths and weaknesses, autocratic/democratic urges and so on. The safety manager should encourage:

- the expression of positive and negative feelings;
- people to stay with transactions;
- structured exercises that cause individuals to be more aware of what they want from others;
- people to push towards greater authenticity for everyone.

Although these exercises may be carried out in groups the focus is usually on the individual members of each group. There is a danger that some staff could feel that they are being coerced into some sort of therapeutic situation if not used correctly. In order to avoid this problem it is recommended that Gestalt exercises only be conducted by staff who are properly trained in this area of specialisation.

Team building, however, can produce extremely powerful results. The reason for this is that the intervention process is in harmony with the nature of an organisation as a social system. For example, under a system involving the division of labour, some aspects of the total organisational tasks are assigned to teams. Team or group assignments are then broken down and assigned to individuals. More often than not, individuals within such a team relate independently to each other and should integrate and co-ordinate individual efforts in order to achieve success in task performance.

Some particularly important issues to consider within the safety system are:

- communication;
- the roles of individuals and groups;
- group problem solving and decision making;
- group norms and growth;
- leadership and authority;
- intergroup co-operation.

GROUP BEHAVIOUR

Behavioural scientists have increasingly come to regard the study of groups within organisations as a profitable and important area. They have attempted to answer the following questions:

- What constitutes a group?
- How important are group processes in determining organisational effectiveness?

The answers to these questions are usually based on an understanding of the dynamics or processes that take place within groups. For example, the interactions of behaviour between people, status, authority and leadership patterns, the relationship between physical behaviour and the emotional climate or feelings that exist within the group. It is felt that an understanding of group dynamics is best obtained using practical rather than theoretical means. However, some basic concepts which should be understood are discussed below. These include an understanding of the relationship between a work group and its organisational environment and the notion of behaviourial norms.

Behavioural scientists define a group as any number of people who:

- interact with one another;
- are psychologically aware of one another;
- perceive themselves to be a group.

The second and third concepts are important in that they give a means of distinguishing between potentially similar situations. For example, which of the following fits the above definition in all respects?

- the staff in the safety department;
- a football team;
- the passengers on a hijacked aircraft;
- a church congregation;
- a fan club.

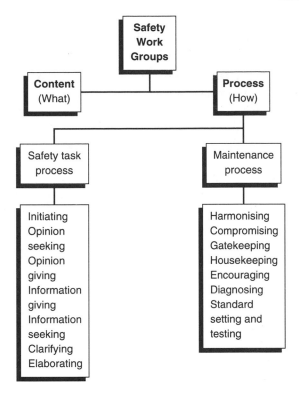

Figure 3.3 Conceptual model of the working group

The definition also focuses attention on some of the areas concerned with working group effectiveness. For example, the way members interact, the psychological and emotional processes within the group and the group's feelings of cohesion and unity. The following model is useful in understanding groups.

Conceptual model.
Group tasks can be split into two types. These are:

1. Content needs: what the group has to do. This will require certain types of resource.
2. Process needs: how is the group to operate? What process will it go through in order to meet its objectives? Process needs can be further split into the following:
   (a) task process: doing things primarily concerned with solving the problems in an effective and constructive way; and

    (b)   maintenance process: those activities which are concerned with producing feelings of cohesion and morale amongst group members almost irrespective of the particular task being undertaken.

These needs are illustrated in Fig. 3.3, and some examples of the task process and maintenance process activities are given.

It will be noted that the examples of process behaviour given in Fig. 3.3 are all positive types of behaviour which are felt to contribute towards effective group working. Risk is listed amongst those negative factors which can hinder a group's behavioural effectiveness. The same piece of behaviour can and often does meet content and process needs at the same time. For example, a good idea to reduce accidents expressed at an appropriate moment might shift the group on towards its objective and generate feelings of intense satisfaction amongst group members at the same time.

Effective working group

The model suggests that an effective working group can be viewed as one where both content and process needs are identified and agreed as appropriate rather than where too much emphasis is placed on either content or process-orientated behaviour. This requires a flexible role structure similar to that shown in Fig. 3.4.

An effective working group is seen as one which is characterised by:

- higher quality decisions;
- more appropriate leadership;

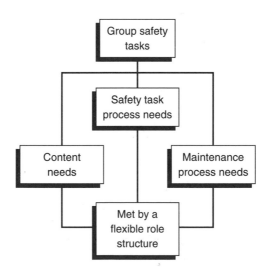

Figure 3.4   Flexible safety role structure

- distribution of authority according to task needs;
- high level of trust and openness leading to co-operation rather than competition between group members.

Some aspects of group behaviour

Solidarity is a key issue within a group context and without it the group would cease to exist. Other issues are:

- Similarity: the more the group members have in common, the greater the solidarity. If everything else is equal there will be greater solidarity within single sex or single race groups than within mixed ones.
- Pressure for conformity: the more willing a group is to exclude or punish a member the stronger the bonds that keep them together become, thus the more intolerant the group becomes, the greater its power.
- Equality: the greater the equality between members, the greater the solidarity. This means that a social group can be weakened if someone deliberately sets out to make members unequal, for example using awards or medals to neutralise informal opposition.
- Safety issue: the bigger the issue is perceived to be, the greater the solidarity induced. This is where opinion leaders are crucial in building up the significance of safety issues.
- Membership changes: no social groups can survive large and frequent changes in membership and continue to be effective.
- Participation: the opinion leader who makes judgments without consulting his group helps to destroy its solidarity.
- Dependence: the more the members depend upon each other, the greater the solidarity. In such circumstances a threat to one becomes a threat to all.
- Interaction: the more often the members of a group interact with each other (i.e. meet, chat, signal and share), the higher the solidarity is likely to be. Interactions create common attitudes and common responses.

Within groups there are two types of leader. The informal leader owes his position in the group to image power in that the group recognises his competence and identifies with him. The formal leader's identification and competence are important, but so is the legitimacy of his role and this is not decided by the group, but by the organisation as a whole. In a formal group people know when they are leaders, but in informal situations people are never certain what their role status really is at any given time.

From the safety manager's point of view, some possible roles in groups might be:

- The professional: normally less militant than the activist except when his job is threatened. Can suffer complacency and delusions of invulnerability with time.

- The activist: does it for love! This person has to be prepared to take their own decisions, own punishments and own personal satisfaction in what is done. If not, this person could become a professional.

- The playing member: if too much is asked of this person he may become a non-playing member or leave if pushed too far.

- The non-playing member: usually takes the benefits provided and justification for the action. In some safety committee situations he has to pay his dues in which case he has the option of being a licensed critic as well (it is the outsider who is the unlicensed critic).

- The outsider: keeps himself indignant and fascinated with issues under discussion. The unlicensed critic lives through the experience of others. Tolerance, dignity and humour are usually the best responses to outsiders.

- The spectator: here the main preoccupation is keeping amused and concentrating on variety, excitement and changes in lively people with flitting minds.

The group and its environment

The complexity and nature of work organisations means that employees often find themselves members of teams or groups. Few people work entirely on their own and their behaviour is, as a result, influenced by the nature of the groups of which they are members. Experience tells us that when people work together, they develop ways of thinking and behaving which are characteristic and not strictly necessary in order to perform the task for which the group came together. Following on from this is a description of a particular conceptual model or systems model which has proved useful in understanding how various elements come together to generate the behaviour that is characteristic of a particular working group.

This model is concerned with recognising the difference and the relationship between two types of behaviour:

- The behaviour with which the group begins. Some behaviour is expected by external factors in the organisational environment such as the task the group is given or communication channels to be used. This is referred to as the required behaviour. Some behaviour is determined by the nature of the members of the group such as skills possessed or work habits learned previously. This is referred to as given behaviour.

- Behaviour which develops internally, which is greater than that required or given e.g. the way in which some groups set themselves higher output levels than others. This is referred to as emergent behaviour.

The relationship between the prescribed and the emergent aspects of a work group's behaviour is considered to be an important element in understanding the functioning of organisations. The model uses three categories of elements of behaviour:

1. Activities: what people do such as running, walking, talking, sleeping, eating, operating machinery, etc.

2. Interactions: communications or contacts between people in order to relate the activity of one person to the activity of another e.g. a conversation or one person handing something to another.

3. Sentiments: an idea, feeling or belief about the work or about the others involved e.g. 'making cars is a good thing'.

These three elements are combined with the notions of required, given and emergent behaviour and are used in the manner illustrated in Fig. 3.5.

Background factors or the organisational environment of the group under study will determine the nature of the required and given behaviour. This might be thought of as the starting point in behavioural terms for that group. As a consequence of the group being in existence, it will develop its own characteristic emergent behaviour, in addition to that which is formally prescribed. The total pattern of behaviour will have consequences in terms of productivity, satisfaction and individual development. These will in turn feed back and cause modifications in the required behaviour for that group.

Issues in work group behaviour and norms

The following are considered the important background factors in determining how a work group should develop:

- Is emergent behaviour good for the organisation?
- Can effective teams be built?
- How important are group norms in determining effectiveness?

One of the most frequently observed characteristics of the work group is the way in which it quickly develops a culture of its own. A particular work group or team is thought of as having an identity of its own, and the behaviour of the members as predictable. The major feature of this group culture is the development of norms which are standards or codes of behaviour to which members of that group conform. Safety practitioners might develop norms about conduct, dress, language or perhaps who does what, how, when, etc.

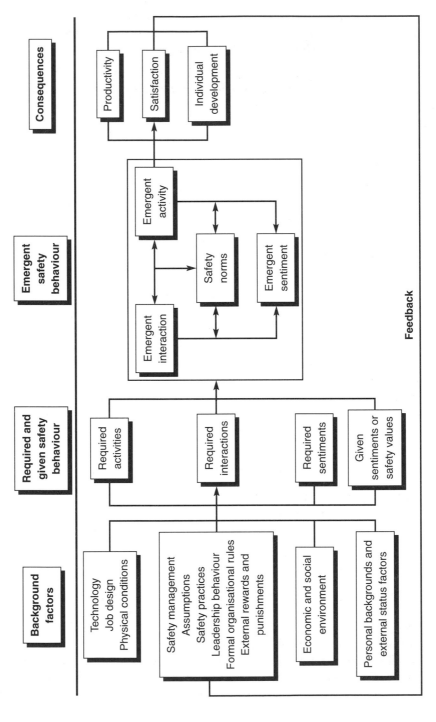

Figure 3.5 Elements of group behaviour

The kind of norms mentioned might be thought of as trivial, but the desire to conform seems to extend to areas which are more important from the work organisation's standpoint. For example, early experiments have shown that group members were very often willing to modify their judgments in order to conform to what they thought other members of the group might accept. Researchers attributed this phenomenon to an attitude of social conformity that people adopt in social situations. If such conformity is widespread, then it will have implications for team building and exploiting the creativity of individuals within a work group. Another example of the operation of norms is when output amongst members of a production team is restricted and becomes broadly similar, even when the members work independently or on individual incentive schemes. This phenomenon was first examined in some studies referred to as the Hawthorne studies conducted in the 1920s but would appear to have bothered work study officers ever since! Norms do not only affect lesser issues such as language and dress but also factors such as output, creativity, resistance to change and so on.

Three propositions

The group is a powerful instrument for influencing or controlling its members, but it is clear that groups differ markedly in the extent of such influence. In order to understand the problems of conformity of members within groups, three propositions are offered:

1. The more attractive a group is to its members, the more likely members are to change their views to conform with those of others in the group.

2. If an individual fails to conform, the group is likely to reject him; the more attractive the group is to its members the more decisively they will reject this individual.

3. Members are more likely to be rejected for deviancy on an issue that is important to a group than on an issue that is unimportant in that non-conformity on peripheral issues is tolerated.

INDIVIDUAL ROLE ANALYSIS

It is not possible, within the context of this chapter, to explain and describe the nature of the individual in some absolute sense; nor is it possible to examine here all the aspects connected with personality, motivation and learning or cognitive factors such as intelligence although these areas are considered later in Chapters 8 and 10. As far as human resources management is concerned with safety, two areas of study have been selected, relating to the individual, which appear to be important in the understanding of behaviour at work. The first of these is motivation and the second is perception and how the individual views the environment about him. However, there are two questions about human behaviour that are important to the safety manager.

In what way can we regard people as being the same?

From an examination of current notions about human behaviour, it can be seen that three basic concepts are supported by most psychologists. These are:

1. Causality: human behaviour is caused by heredity or by environmental factors acting on an individual.
2. Directedness: the notion that people desire things in such a way that their behaviour is directed towards some goal.
3. Motivation: goal-orientated behaviour which has an element of intensity behind it usually indicated by words such as want, need or drive.

These three ideas can be linked together as a closed system in order to form a simple model for understanding human behaviours. Such a model is given in Fig. 3.6.

An individual receives some form of stimulus, for example, a headache. When this stimulus becomes severe enough, motivation is raised to a level whereby some form of action is pursued which will enable the goal of relief. Two aspirin tablets are swallowed thus achieving the goal and in so doing removing the stimulus, causing this particular piece of behaviour to stop. In such a case there are a number of issues raised by the adoption of this particular model:

- What about the notion of free will?
- Why take aspirin rather than paracetamol?
- Some goals do not seem to be achieved totally (the stimulus is not removed completely). What about goals such as ambition and status?

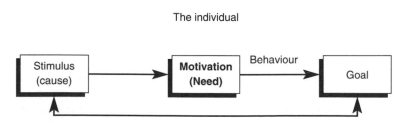

Figure 3.6 A basic model for understanding human behaviours

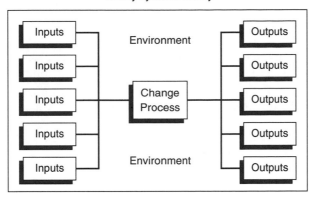

Safety systems theory

Figure 3.7 A model of the change processes

People are the same in that their behaviour is caused, goal directed and motivated and that they have roughly the same sort of physical features to operate with.

In what way should people be regarded as being different?

People can be regarded as being different in the way that they approach and operate the model illustrated in Fig. 3.7. Any individual will develop or possess particular personality characteristics.

Here, personality is defined as the pattern of relatively stable states and characteristics that will influence an individual's behaviour in the achievement of goals. Personality is made up of many components and it can be recognised in the importance of individual differences in:

- The abilities and aptitudes an individual has or develops.
- The attitudes possessed or developed.
- The pattern of need or motives maintained by the individual.

Developing individual needs

From the above, it can be seen that an individual can be classified into two broad areas:

1. Physical;
2. Psychological.

In terms of the physical needs, these are essential to the survival of the animal and include food, water, warmth and sleep. They are easily identifiable and are universal in that everyone has them. Psychological needs, however, tend to be much less easily definable and are much more individualistic. Needs such as prestige, status or ambition seem to exist in some people and not in others. There are also those needs used to make judgments about people. 'He is ambitious' is much more likely to be used than 'he is hungry'. What follows is an attempt to explain how it is that some people have particular patterns of need. Although this is presented in a simple form, it does illustrate the importance of the environment within which individuals exist, operate and learn to behave in certain ways. It is also important to remember that this explanation describes only one approach to the problem.

It is assumed that a new-born baby starts life with:

- His body and physical capabilities which will develop.
- His physical needs.

A characteristic of the human infant is that it is unable to satisfy these physical needs by itself. It is dependent upon others, principally parents, for the satisfaction of its needs. Although the human infant has been described as the perfect example of minority rule, it is the dependency relationship that accounts for the development of the particular pattern of psychological need. To the extent that this dependency produces satisfaction of existing physical needs, feelings are likely to be positive, affectionate and protective. From this it is likely that strong social needs are developed. To the extent that the dependency relationship does not satisfy but rather frustrates existing needs, then to that extent a person is likely to develop feelings of anger and hostility, to wish more strongly for independence and autonomy, and to develop egotistic needs.

The way in which parents are able to meet the dependency needs of their child will depend to a large extent on the general pattern of personality development for that individual child. Teachers, friends, employers, etc., will continue this in later life.

Abilities

The most obvious way in which people can display that they are different is in what they are able to do. Psychologists have studied differences in the areas of:

- Mechanical ability: the understanding of mechanical relationships and the ability to visualise how parts fit together into a whole.

- Mental ability: the intelligence, logical reasoning, verbal and numerical abilities, etc.

- Creative ability: the aesthetic judgment, musical ability, artistic talents etc.

The issue as to whether abilities are inherited or learned is still being studied. More recent approaches take the view that abilities lie on a continuum. At one end are the responses that are geared to physiological capabilities such as dexterity, reaction times, etc., and at the other, those responses where genetic or physical factors are not the restricting factor such as interpersonal skills.

Attitudes

These are learned predispositions to behave towards or to respond to stimuli in particular and predictable ways. For example 'all Jews are good businessmen', 'all Scotsmen are miserly', 'all French women are fashion conscious' or 'wearing safety gear is cissy'! Attitudes can serve a number of purposes:

- They can give order to the way an individual views events and thus help make sense of the constant and varied bombardment of stimuli affecting them.
- Attitudes and values can help a person deal with his or her psychological problems and conflicts.
- They contribute to a person's identity.

There are many attitudes existing in the workplace. Good examples are the two opposing attitudes to work that are held by employees:

1. Work is a means to an end. Whilst it is often unpleasant, it is a necessary evil in order to satisfy other needs.
2. Work is an end in itself. A person gains satisfaction and self-fulfilment through work.

An understanding of attitudes prevailing amongst those being managed will enable the prediction of likely outcomes of alternative safety management strategies. For example, if work is seen as a means to an end, the effectiveness of safety programmes could be limited.

STAGES OF INDIVIDUAL ROLE ANALYSIS

The first stage of the individual role analysis process is concerned with clarifying your line of accountability within the organisation and the essential purpose that your job exists to fulfil.

A clear understanding of job purpose will underpin the subsequent analysis; after all, it is difficult to arrive at valid judgements about performance in the absence of an agreement about what that performance is required to achieve. It also provides a means of assessing the extent to which the activities that absorb your time are really contributing to your primary task.

Job title

The title of your job tends to influence the way in which you and others see your role. As far as is possible, your job title should be an accurate and easily understandable identification of what you do. Consider the following points:

- Is your job title easily understood by those who need to identify you, both inside and outside the organisation?
- Does it accurately describe both your *rank* within the hierarchy and the service *function* you are concerned with?
- Is it a label that you personally are happy to work with?

If your present job title meets all these points then all well and good. If not, then draft an alternative and make a note to discuss this. However, it should be remembered that job titles are a part of the establishment control system of an organisation and your own job title may not be changed without it having wider implications for the company.

Unit or function?

Do you manage a unit, a function or both? A unit is an identifiable part of the organisation, of any size from a section to a department. This means you have direct control over subordinates in your unit. A function is an area of responsibility that extends over people in other managers' units. Your responsibility probably does not give you direct control over these people and you may not have any direct subordinates of your own.

Your immediate manager?

This is the person to whom you are directly accountable (i.e. your 'boss'). Can you identify the one person to whom you are directly accountable for all aspects of your work. What is his job title?

If you can identify more than one person to whom you are accountable, note all their job titles and briefly the nature of the accountability you owe to each of them. Consider whether *multiple subordination* is a problem.

Also:

1. Do you ever suffer from work overload or underload as a result of these reporting arrangements?

2. Do you ever experience conflicting demands from different 'bosses'. If so, whom do you believe to be the person with the responsibility to arbitrate on the conflict or overload? Would everyone else agree?

3. Do the reporting relationships result in aggravation, confusion or stress to you or to others internal or external?

4. Which 'boss' is responsible for your own career and management development?

If you are subjected to *multiple subordination* draft some points on how to overcome this problem.

MAIN PURPOSE OF JOB

This is a brief statement of the reasons for the existence of the job within the organisation. It is necessary to write down the statement in a way which can illustrate what your energies are supposed to achieve by way of service and end product rather than simply summarising the activities themselves. It is helpful to concentrate on the management or supervisory aspects of the responsibilities and ask:

- What is it that I am accountable for when ensuring what happens as a result of utilising resources over which I have been given authority as a manager?

As a general rule the main purpose statement should only include one verb since to have more than one implies that the statement does not describe the *single main purpose* of the job.

Your immediate manager's job purpose

Draft a statement of your immediate manager's main job purpose for discussion and consider the following:

- Is it essentially the same as yours?
- Is it the same but for a larger geographical area?
- Does it cover a number of functions of which yours is one integral part?
- Is it unrelated to yours?

What are your responsibilities

The precise delineation of responsibility is probably the most difficult part of drawing up a job description; in fact, many job descriptions omit responsibility altogether. Confusion about responsibilities is a frequent cause of conflict between 'boss' and subordinate and of delays in decision making.

Part of the confusion about responsibilities is due to the various meanings of the word *responsibility*. To avoid this, two words are used to describe the two different aspects of responsibility. These are accountability and authority. It is important that the two words are clearly understood.

ACCOUNTABILITY

This is the process by which someone is held responsible for something which has been delegated and not relinquished. The person who actually takes a decision is not usually the person accountable to the workforce or public (or shareholders) to whom account must ultimately be made. He should be accountable only to his immediate manager, who in turn is accountable to his manager and so on. Thus within any chain of command accountability may be owed for a single item at many different levels of the chain at each boss-subordinate link.

AUTHORITY

This is the counterpart of accountability and it is regarded as the right to influence a decision. Authority is the process by which you are given responsibility. The influencing of decisions may be by degrees but there can also be a right to advise or make recommendations about a decision, and also a right to make a personal decision without referring to a superior first.

Authority, unlike accountability, is relinquished when it is delegated. It can, of course, be withdrawn but then it is no longer delegated. Like accountability, authority passes along the chain of command and is delegated authority from above.

The definition of your formal authority is of great importance in clarifying your role as a manager. Many managers complain, however, that not enough authority has been delegated to them or that it is not clear how much delegation has taken place.

In theory, you can only be held accountable for resources to the extent that you have authority to decide about them. For example, if you are not entitled to appoint your own subordinates you should not be held responsible for the results of poor selection. In practice, the line is often very difficult to draw, particularly as delegated authority is always limited in some way. For instance, managers are usually only authorised to appoint subordinates up to a certain level, often have to consult others in the process and must adhere to general procedures either laid down by the Personnel Department or within a certain establishment.

It is important that accountability and authority are correlated as far as possible and that every individual knows just what the limits of their formal authority are. Consider the following:

- If you do not have all the authority you would want, where do the restrictions come from? From external restrictions placed on the organisation or from policies within the organisation?
- Who possesses the authority that you would like? If the authority resides with a group, panel or committee what is your role within that group?

Some areas of accountability and authority

- Accountability for staff;
- Authority over staff;
- Accountability for finance (income and expenditure);
- Authority over finance;
- Accountability for physical resources;
- Authority over physical resources.

AUTHORITY-ACCOUNTABILITY REVIEWS

List any discrepancies between the accountability you have to provide a service or produce results and the authority you have been given to do so. Draft any changes you feel appropriate.

Some questions to consider are:

- What do we want?
  - improved communication skills;
  - better methods of task implementation;
  - to know where safety appraisal fits within the department;
  - to know how to criticise and to encourage;
  - to appear professional;
  - to be able to recognise strengths and weaknesses;
  - to know how to deal with the questions listed here.

- What are our concerns?
  - what does my company want from me?

  - safety appraisal of reluctant staff;
  - safety appraisal of staff who have more knowledge than me;
  - as time goes by, the problem dies away;
  - what are the right answers;
  - staff time – where do we find it?
  - being appraised by someone who does not know enough about it;
  - constant changes in work loads;
  - need for care, help and trust;
  - the need for safety appraisal;
  - breaches of safety could become disciplinary matters.

- Key result areas?
  - are they now a distinct part of my department?
  - where and how do I take the decisions?
  - attempts to be successful;
  - success = safety performance standards
  - adequate safety budget;
  - no overspend;
  - spend within x per cent of forecast;
  - time....by the agreed date;
  - safety....auditors find no errors;
  - cost of producing safety information is reducing by x per cent per....;
  - quantity = accident reduction targets;
  - quality.......

- What do we believe is important?
  - risk is reduced to a minimum;
  - accident targets are achieved;
  - dangerous occurrences are controlled;
  - trust amongst workforce;
  - openness and honesty;
  - constructive criticism is given;
  - safety performance interviews are relaxed;
  - ground rules are clear;
  - confidentiality;
  - idea of a contract is strong;
  - preparation;
  - action for change must be agreed;
  - background information must be available;
  - respect for other's point of view;
  - safety appraisal interviews are not interrupted;
  - layout puts people at ease;
  - both parties to be ultimately agreeable.

- Pre-safety appraisal
  - performance standards – where are we now?
  - agreement to safety objectives for the main appraisal;
  - allow time to prepare reasons for success or failure;
  - agree what information is required;
  - monitoring of the action plan;
  - outline main purposes of the safety appraisal sessions;
  - monitoring of plans to achieve safety performance standards;
  - statement of the areas for discussion;
  - Examination of KRA's;
  - Outline what is to be written down or verbally passed on.

- Important points for individuals
  - help on how to measure individual performance;
  - training on how safety performance fits into company policy;
  - recognise when help is needed;
  - background information to safety performance management;
  - safety appraisal should spend 80 per cent looking forward optimistically;
  - safety performance management is to lower risk and reduce accidents;
  - where, when and how problems can be highlighted;
  - that satisfactory safety appraisals are correctly communicated;
  - how to encourage constructive criticism.

- Safety performance standards
  - they must be measurable;
  - include how the job holder sees success with the Key Result Areas (KRA);
  - Is success in your KRA to do with:
    accident reduction
    accident prevention
    lowering of risk
    behavioural improvements
    time
    quality
    costs
    quantity
    talking
    doing
    all, some of these or additional safety performance standards;
  - agreement on safety performance standards which have not been met;
  - agreed written action plan;
  - equal contribution rate;
  - praise given where it is due;
  - agreement of safety performance standards to be improved upon.

SUMMARY

It is important that everyone involved in safety auditing is aware of his role and function. It is these aspects that would also be included as part of the safety auditing process. All too often safety auditing is undertaken independently without the direct involvement of the workforce. This is not a recommended practice because research now shows that the involvement of all staff and the complete understanding of one anothers' roles and functions in the safety process is an essential feature of successful safety performance appraisal. At the same time some safety auditing processes only include the recording of pre-determined areas in order to examine the effectiveness of safety policies and programmes and included in some of these are methods of awarding points or issuing scores. These should be avoided because they can give management a false sense of security and there have been cases of recently safety audited organisations obtaining figures of approximately 90 per cent at safety audit time during one week only to be faced with a major disaster the week after!

It is far more effective for the safety manager systematically to go through the information gathering process first. It is essential that safety performance standards are agreed by everyone involved and that they are, wherever possible, clearly quantifiable. Everyone must be aware of their role within the overall company safety policy before other aspects of the safety auditing process can be considered. A further aspect to be considered in the information gathering process is that the safety manager should gain an understanding of task analysis techniques so that an understanding and appreciation of each of the tasks which will be subjected to safety audit is obtained. This is discussed further in the next chapter.

4 TASK ANALYSIS

In this chapter we will look at the following:

- **Introduction to task analysis**
- **Task identification and questionnaires**
- **Status and identification**
- **Duties and responsibilities**
- **Human characteristics and working conditions**
- **Safety performance standards**
- **Data collection**
- **Contribution of operations management**
- **Problem solving strategies**
- **Operational safety management issues**
- **Work measurement**
- **Job evaluation**
- **Job descriptions and specifications**
- **Critical path analysis (CPA)**
- **Summary**

INTRODUCTION TO TASK ANALYSIS

In order to have a proactive approach to human resource information, it is necessary to be aware of not only the environmental challenges evident in the working environment but also the equal opportunities which are considered the cornerstone of human resource information systems. Included in this are the planning and information about various jobs and their specific tasks which are central to organisational productivity and safety output. If task analysis is carried out correctly, the organisation is more likely to meet its objectives. If not, productivity and safety suffer and the company is less likely to meet the demands of society, customers, employees, managers and others who have a stake in its success. An illustration of the human resource information system in relation to task analysis is given in Fig. 4.1 below.

Task analysis initially requires some basic information about the characteristics of the job, standards, safety and human abilities required for the jobs that are carried out. As personnel science has grown, the scope and complexity for job analysis information is systematically required and this is

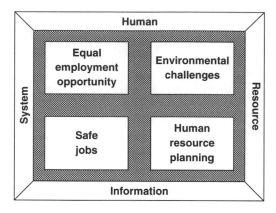

Figure 4.1 Human resource information system

important for the safety audit. Task analysis collects, evaluates and organises information about jobs. The actions may be carried out by specialists or by the safety manager who is interested in specific areas of the task detail in order to assess risk. Some safety management actions which rely upon job analysis information are given in Fig. 4.2.

Carefully recorded task information is crucial because it affects so many activities. Some of the affected areas are given in Fig. 4.2, but without task analysis information, it is difficult to identify the environmental challenges or specific task requirements which would affect the quality of work life. In

Action	Comment	Check
Evaluate	how environmental challenges affect individual jobs	
Eliminate	unnecessary job requirements which can cause discrimination	
Discover	task elements that can reduce/increase risk	
Plan	for future human resource needs	
Ensure	right people for right jobs	
Determine	safety training needs for new and experienced employees	
Create	opportunities to develop employee safety potential	
Set	realistic safety performance standards	
Place	employees in jobs that use their skills effectively	
Compensate	job holders fairly	

Figure 4.2 Some safety management actions which rely on job analysis information

order to match job openings with job opportunities it is necessary to have a detailed understanding of what each job entails. Indeed, in order to be appreciative of the risk involved it is also necessary for the safety manager to be fully aware of the task detail information.

Task analysis specialists gather information about jobs and their specific tasks. In general terms they collect information about:

- job purpose;
- job design;
- task inputs (people, materials and procedures);
- task outputs (products and services);
- company safety policy;
- other reports about how the work should be carried out;
- the task analysis techniques to be employed.

This will enable:

- the identification of jobs to be analysed;
- the development of a task analysis questionnaire;
- the collection of the correct information.

Task identification and questionnaires

It is important to identify the different jobs and tasks in the company before the information is collected. Where task analysis has been carried out before, the safety auditor will require this information before the safety audit commences.

In order to study the detail involved in jobs, it is necessary to develop a checklist or questionnaire. These are sometimes referred to as job analysis schedules. Regardless of what they are called, these forms seek uniformly to gather information about each task. The intention of the questionnaire is to discover the duties, responsibilities, human abilities required, performance standards and safety standards of the tasks investigated. It is important to use the same questionnaire on similar task analysis exercises on similar jobs. You should seek information which reflects differences in task rather than differences in questions asked.

Some typical questions which are asked at this stage are given in Fig. 4.3. The main parts of this are discussed below.

Status and identification

The first two headings in Fig. 4.3 indicate how current the information is and the identification of the job described. Without these entries, users of task analysis data may have to rely on out of date information or worse,

Bloggs Construction Ltd

Job analysis questionnaire

Part 1 Job analysis status:	
Job analysis form revised on	
Previous revisions were made on	
Date that job analysis was conducted	
Name of person carrying out job analysis	
Verified by	

Part 2 Job indentification:

Job title		Other title	
Division		Department	
Section		Supervisor(s) title	

Part 3 Job summary:

Briefly describe the purpose of the job, what is done and how is it done

Part 4 Duties:

The primary duties of this job are best described as:

Technical		Managerial	
Administrative		Clerical	
Professional		Labouring	

List the major duties of this job and the proportion of time spent on each			
Major task 1		% time	
Major task 2		% time	
Major task 3		% time	
Major task 4		% time	

Figure 4.3 Some questions about job analysis

Bloggs Construction Ltd		Page 2

Job analysis questionnaire

List the major duties of this job and the proportion of time spent on each			
Task 1		% time	
Task 2		% time	
Task 3		% time	
Task 4		% time	

State what criteria constitute successful safety performance of these duties

How much safety training is required to meet these safety performance standards?

Part 5 Responsibilities:

What are the responsibilities found in this job and how signficant are they?

Responsibility for	Significance of responsibility	
	Minor	Major
Accident investigation		
Accident analysis		
Equipment operation		
Equipment maintenance		
Material usage		
Protection of equipment		
Protection of tools		
Protection of materials		
Personal safety		
Safety of others		
Other's work performance		
Other's safety performance standards		
Safety clothing		
Wearing of safety clothing		
Issue of safety clothing		
Maintenance of safety clothing/equipment		
Other activities (please specify)		

Figure 4.3 cont. Some questions about job analysis

Bloggs Construction Ltd | Page 3

Job analysis questionnaire

Part 6 Human characteristics:

What physical attributes are necessary to perform the job safely?

Of the following characteristics, which ones are required and how important are they?

Characteristic	Not required	Useful	Essential
Eyesight			
Hearing			
Speaking			
Sense of smell			
Sence of touch			
Sense of taste			
Hand-eye co-ordination			
General co-ordination			
Strength			
Height			
Health			
Initiative			
Integrity			
Judgement			
Attention			
Ability to read			
Numeracy			
Writing			
Education level			
Ability to work at heights			
Ability to work underground			
Experience			
Other (please specify)			

What safety training is required to cope with deficiencies in experience?

Figure 4.3 cont. Some questions about job analysis

Bloggs Construction Ltd Page 4

Job analysis questionnaire

Part 7 Working conditions:

Describe the physical conditions under which the various tasks are performed

What are the psychological demands of the job?

Describe any unique features of job safety performance

Part 8 Specific health and safety requirements:

Describe in detail all hazardous conditions associated with this job

List specialist safety training or equipment which is required for this job

Figure 4.3 cont. Some questions about job analysis

Bloggs Construction Ltd   Page 5

Job analysis questionnaire

Part 9 Safety performance standards:

How is the safety performance of this job measured?

What identifiable factors most contribute to the successful safety performance of job?

Part 10 Additional comments:

Are there any points which should be noted?

Signature of job assessor

Print name of assessor

Date

Figure 4.3 cont. Some questions about job analysis

apply it to the wrong job. Since most jobs do change over a period of time, outdated information may misdirect other task analysis activities.

Duties and responsibilities

Many forms briefly explain the purpose of the job, what duties are performed and how they are performed. The summary provides a quick overview of the job. The specific duties and responsibilities are listed to give a more detailed insight into the position. Questions concerning responsibility can be explained when the checklist is applied to management jobs. Additional questions map areas of responsibility for decision making, controlling, planning, safety and organising and other management functions.

Human characteristics and working conditions

In addition to information concerning the job, it is important to obtain details of the qualifications required to perform the job safely. This area of the checklist uncovers the relevant knowledge, skills, abilities, training, education, experience and other characteristics that job holders should possess. Working conditions may explain the need for particular skills, training, knowledge or even job design. Knowledge of hazards allows for the redesign of the job or the protection of workers via a number of safety management measures (e.g. safety training or the provision of safety equipment). Differences in working conditions influence hiring, placement and compensation decisions.

Performance standards

The job analysis questionnaire must seek information about job standards and safety which will be used for the evaluation of performance (particularly safety performance). The information is gathered on jobs with clear objective standards of safety performance. Where standards are not readily apparent then it is necessary to seek advice from supervisors and managers about the setting of quantifiable safety performance standards.

DATA COLLECTION

There is no best way of collecting all the information necessary for job and task analysis. One must evaluate the trade-offs between time, cost and accuracy associated with the use of interviews, juries of experts, questionnaires, employee logbooks or some other combination of these techniques.

Interviews

These are an effective means of collecting job analysis information. The interviewer will have the job checklist as a guide but can add other questions where appropriate. Although this process is rather slow and expensive, it does allow the interviewer to explain unclear or ambiguous questions and to probe further into uncertain answers. Both job holder and supervisor are interviewed in this way but the auditor usually talks to a number of other workers first. The information obtained from the supervisor and job holder will then verify the data collected. This pattern ensures a high level of accuracy.

Jury of experts

These can be rather expensive and time consuming but are essential for some very important jobs. The jury usually consists of senior job incumbents and immediate supervisors. Together the group represents considerable knowledge and experience about the particular job and its tasks. In order to get the job analysis information, interviews are held with the group and the interaction of the members during the interview can and does add insight and detail to the discussions. A benefit of this process can be the clarification of expected job duties among workers and supervisors who are on the jury.

Mail questionnaires

This is a quick and much less costly exercise than the two previous methods. This approach does permit several jobs to be investigated at once and at a small cost. However, there is less accuracy because some questions can be misunderstood, or incomplete and there may be unreturned questionnaires. The advantage of this method is that supervisors can be given the same questionnaires so that responses can be verified.

Employee log

An alternative is where workers keep activity log books or diaries. Workers usually keep a record summary of their jobs and tasks in the log books. Where entries are made over the entire job cycle the log book can be quite accurate. It may even be the only way to collect job and task information particularly when interviews, consultants and questionnaires are unlikely to assess the entire job.

However, log books are not a popular technique because they are time consuming and this can make them rather costly. Some managers and workers actually regard them as a nuisance and as a result resist their use. Once the novelty has worn off, accuracy can also decline as entries become less frequent.

Observation

This involves direct studies of job holders through observation exercises. It is slow, costly and potentially less accurate than other methods. Accuracy may be rather low because the observers may miss irregularities in task activities but this method is preferred in some situations. It is particularly important when data gathered from other techniques, observation may confirm or remove doubts about a particular problem. Language barriers can cause observation to be used especially in developing countries or where a large number of foreign language workers are employed.

Combinations

Since each of the methods described above has faults, it may be prudent to employ more than one of these techniques concurrently. Such combinations can ensure high accuracy at a reasonable cost.

CONTRIBUTION OF OPERATIONS MANAGEMENT

There are three types of approach to the subject of operation management used by practitioners in industry. Traditionalists tend to concentrate upon the clerical procedures of 'production' type management and use work study, value analysis, standardisation, production control and other similar techniques emphasising the qualitative rather than quantitative aspects of management. Such practitioners lean towards synthesis rather than analysis as their main term of reference. A further type rely mainly on mathematics and the applications of operational research and statistics to management analysis whilst the third type uses a mix of the first two. This type is the one which shall be discussed here.

For our purposes, production function concerns two types of decision. Firstly, those pertaining to the design or establishment of the system, and, secondly, those concerning the operation, performance and running of the system. This philosophy refers to the safety management process and the establishment of the safety management system and its performance. A system is not only an arrangement of physical facilities but also the predetermined manner in which safety related strategies are made and implemented. Decisions relating to system design include such things as the acquisition and arrangement of equipment, determination of safe manufacturing or operational methods whilst decisions relating to the operation of the 'production' process concern the implementation, monitoring and evaluation of safety strategies. These can be regarded in some cases as short term in nature.

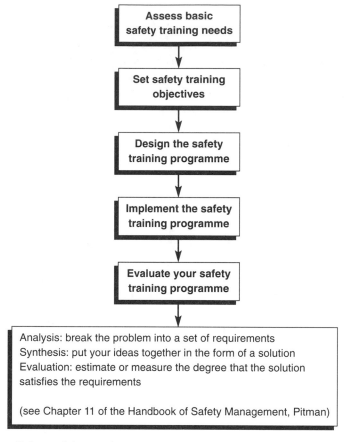

Figure 4.4 Safety training needs

Operations management is concerned primarily with the design and operation of efficient and effective systems whose purpose is the provision of goods or, in our case, safety services. Applying the basic principles to the safety training function is shown in Fig. 4.4.

Solving problems

When a safety problem has been identified systematic search is often less time consuming, cheaper and offers easier implementation of the remedial strategy/ies than a series of plausible guesses which prove fallacious. This is shown diagrammatically in Fig. 4.5.

The design of jobs has usually been the responsibility of the production function, and effective methods of doing work are essential for the efficient

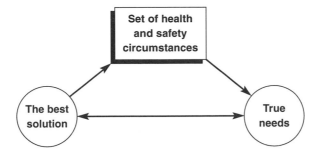

Figure 4.5 Safety problem solving

design of safety systems. The development and implementation of optimum work methods frequently requires the study and design of workplaces and equipment. Control is achieved by performance standards which are obtained after exhaustive investigations into work methods. Such systems are relevant to the way in which the safety manager may set about the provision of the safety service within his work environment as well as to the safety factors built into the general production process for the workforce in each respective industry.

The traditional frameworks of work study are as follows:

- Select the situation to be studied and outline the purpose of the exercise.
- Record the relevant data, including past records and any specialist records.
- Analyse the accident and dangerous occurrence data critically.
- Develop the most acceptable solution.
- Install the new system.
- Maintain the new system by control procedures.

The general model used within a work study framework requires the following phases to be considered.

Diagnostic phase

- Clarify the problem.
- Set objectives.
- Establish the current situation.
- Examine internal and external factors.

Assessing the need to change:

- Identify areas requiring change.
- Highlight financial and accident reduction advantages.
- Identify objectives for change.

Developing alternatives:

- Assess the degree to which each alternative meets the need and/or objective.
- Evaluate financial and accident prevention savings.
- Conduct feasibility exercise(s).
- Select the 'best' solution.

Recommendation:

- Presentation and selling the chosen changes.

Implementation:

- Actioning change (communication needs).
- Gaining support and commitment for the change.
- Debugging the new systems.
- Provision of training and development for people.

Safety performance review:

- Measuring the degree of achievement.

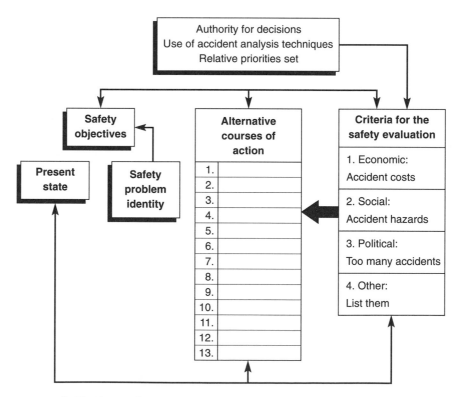

Figure 4.6 Clarifying safety problems

- Noting side effects and unexpected problems or successes.
- Intervening to change objectives or system operation.

Corporate appraisal

This requires the safety manager to consider the implications of any revised strategy and the effects this will have upon the internal environment. There are many theories relating to this process but for practical purposes the following list of questions will provide a framework on which to proceed:

- Where are we now and why?
- What are the strengths and weaknesses of the safety situation under study?
- What are our objectives? (What should they be?)
- Do we need change? (Threats?)
- What alternatives are open? (Opportunities?)
- What does each alternative offer?
- Which alternatives should be chosen?

A general model showing the clarification process is given in Fig. 4.6.

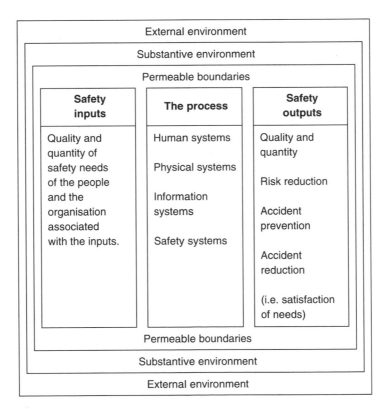

Figure 4.7 Safety systems approach to solving safety problems

Systems approach

This sees all inputs into a safety system, safety organisation or safety process as associated with a set of needs. The degree to which these needs are being met by the process operation will affect the entry back into the system for future inputs (see Fig. 4.7). Also, since the outputs permeate both the substantive and the external environment, they will also affect ability to attract new types of input.

Regulation can only be affected by changes in the inputs or process operation. Choice of inputs is limited by demand for entry again affected by outputs. Process change may be possible with no change in inputs.

The approach aims to identify inconsistencies between the various elements in the model. For example, discrepancies in satisfaction of input needs by outputs and the interactive effect between factors in both substantive and external environment on inputs.

Characteristics of problem solving

There are seven basic characteristics to consider when problem solving.

1. Cycling and recycling between stages takes place until the safety manager has found an acceptable solution to the problem.

2. The amount of search is related to uncertainty at the start. A poor strategy is one which leaves the safety practitioner solving the problem still uncertain which information he cannot use.

3. There will be several paths leading to a satisfactory conclusion. However, they are unlikely to be equal in terms of man hours involved, cost, information needs or organisational disturbance.

4. The chosen safety strategy for problem solving should force the safety practitioner to pose and answer the questions which express the uncertainty with which he begins.

5. To avoid any errors in the final conclusion, the safety manager has to balance the reliability of his information against the cost of verifying it. Sensitivity testing is an integral part of the problem solving process.

6. A good strategy places no constraint on the freedom to change plans in the light of new information and insights as the investigation proceeds. However, at any point in time, one should have a particular plan to which one is working. New ideas can be kept separate until the safety practitioner feels it appropriate to review the results of the current investigation and decide whether to continue on the existing path or to develop a new plan for the next part of the investigation.

7. Throughout the problem solving process, value judgements will affect the conclusions drawn and actions taken. Evidence can never give certainty to predicted outcomes but merely affect the confidence we have in the particular conclusion and the error margin of the predicted outcomes.

Problem solving dilemmas
There are three basic dilemmas to consider:

1. Which goals or objectives?
2. Which strategy or approach?
3. How to overcome a barrier?

These are now discussed in more detail.

Which goal or objectives?
When a number of different goals present themselves, it is often difficult to decide which goal to go for. It is common for any safety problem to be tackled in several different ways. Some dangers are:

- Choosing too broad a field of investigation.
- Goals which are incompatible in respect of problem solving methods.
- Human change needs.
- Data requirements.
- Development needs of the individual.
- Goals impossible to achieve in the time or with the resources available.

Which strategy or approach?
Some dangers to look out for are:

- Choosing an approach which avoids bringing human problems into the open.
- Following an approach used before for a similar problem and finding it does not succeed.
- Failing to assess the readiness for change and the human environment before making the choice.
- Using a non-participative involvement approach.
- Letting the needs for acceptance of recommendation be the sole influence.
- Approaching the problem bottom up rather than from the top downwards.
- Choosing an original research approach rather than a survey of experience of others first.
- Using primary data inputs rather than secondary data.
- Choosing an approach which within the time available leaves inadequate time for analysis and development of alternatives.

How to overcome a barrier?

In following a particular strategy, a barrier can be met which must be sur-mounted before the goal can be achieved. In safety practice these are quite common. Some common barriers are:

- Lack of access to reliable accident or dangerous occurrence data.
- Lack of co-operation or involvement of people.
- Lack of ideas for further development or consideration.
- Obtaining evaluation criteria.
- Obtaining interim decisions or approval affecting further progress.
- Lack of knowledge of the principles, systems, concepts and environment.

Some dangers to consider are:

- Failure or refusal to recognise failure in one's personal approach or behaviour.
- Failure or refusal to recognise excessive change gaps as perceived by others.
- Putting off facing up to the barrier.
- Ignoring the barrier and making assumptions to enable progress.
- Selecting a standard solution rather than use or search for creative techniques.
- Assuming co-operation will come when the final safety recommendations are approved.
- Failure to assess the worth of effort to surmount the barrier.
- Perception of the size and cause of the barrier being significantly different in reality.
- Failure to face up to behavioural problems when involving people outside the safety manager's formal responsibility.
- To accept too readily the impossibility of the situation and give up when it was possible to have overcome the difficulty.

In practice, any combination of these situations may exist. The important thing is to recognise the situation(s) and come to terms with them.

Problem solving strategies

There are three main problem solving strategies for the safety manager to consider:

1. Random strategy.
2. Traditional strategy.
3. Adaptive strategy.

These are now discussed briefly.

Random strategy

In choosing the next step within the problem solving sequence, the safety inves-tigator deliberately adopts a policy whereby the brief is ignored, particularly from previous experience and information. The strategy may be a useful one to

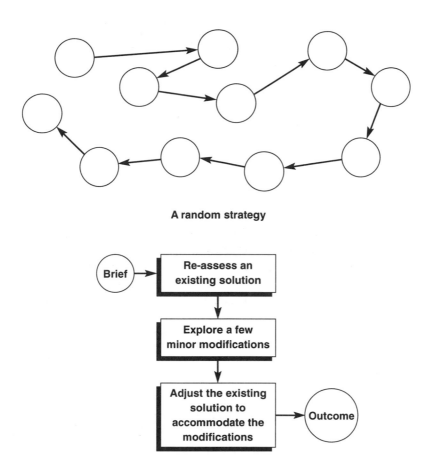

Figure 4.8 *Random strategy and traditional strategy*

adopt in very uncertain circumstances which need to be widely explored before taking up any specific aspect in detail.

Traditional strategy
The safety investigator relies as far as possible on reliable accident and dangerous occurrence information that has been collected for the purpose. A diagram outlining the philosophy of the traditional strategy is shown in Fig. 4.8.

Adaptive strategy
The safety practitioner exploits the information generated at each stage as progress is made between uncertainty and decision as shown in Fig. 4.9. Combining logic and imagination, the results of this adaptive approach are

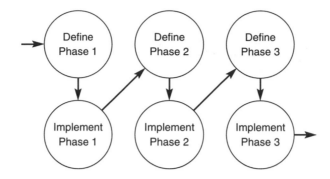

Figure 4.9 An example of a safety adaptive strategy

kept in one set until such time as enough factual and logical information has accumulated in the other set to test their relevance and value. The free thinking data are compared with the sequential thinking data as shown in Fig. 4.9. The next stage will be either to abandon the insight because it does not fit the facts or to abandon the initial line of approach because it can now be seen to lead in an unfruitful direction. When things proceed well, both insights and sequential sequencing point in the same direction.

Personal elements

Effective safety related problem solving is the result of a balance between vision, judgment and knowledge. These will improve with experience. In Fig. 4.10, a diagram is given outlining these three factors and listing those important features within each of the three headings. It should be noted that when applying judgement, one must suppress natural tendencies which may blind one from the truth. A safety practitioner must beware of four areas which researchers have identified as the stereotype syndrome. A safety practitioner should not stereotype problems, solutions, pay-offs or success.

When approaching a problem it is important:

- To question effectiveness before efficiency.

- That symptoms are not causes (identify the cause).

- To establish the relative importance of the various problem areas – consider 'parieto analysis'. This is: the law of the vital few and the trivial many! Chronic problems are those which exist continually. They are likely to give more significant returns if solved than sporadic problems (ones which occur infrequently). Safety practitioners should prevent problems developing into a chronic state.

Understanding	Judgement	Knowledge
* all the related safety systems and related tasks * of changing situations * new requirements * reconciling and recalling previous events * idealising * self-projection into a safety related situation	* truth and relevance of the safety advice and information given * capabilities of people on whom we rely * relevance of action plans and safety auditing processes * demands, priorities and relationships in a situation * significance of facts	* of oneself and others involved * of the situation * of the field of study * of techniques and approaches * procedures, policies, practices and programmes * of relevant theories

Figure 4.10 Some personal elements to consider when safety auditing

- That criteria and evaluation are deemed essential if progress is to be meaningful and controlled.
- To take a system view of the problem situation.
- To define the boundaries of the problem.
- To be prepared to change the approach, boundaries and criteria for evaluation where necessary.
- To consider all the life stages associated with the problem solving situation. For example:
  - the investigation;
  - acceptance/rejection of recommendations;
  - implementation;
  - ongoing operation and maintenance;
  - performance review;
  - growth;
  - modification and change of the safety programme.
- To establish the need for a solution to the problem. For example:
  - to establish a prime need which the safety solution must satisfy;
  - to rank needs according to the importance of them being satisfied;
  - the needs will be associated with all the stages of the problem solving and its resolution.
- To be aware of the hypotheses (unsubstantiated assumptions) upon which the safety approach will be based.

It is important during the analysis phase of the safety investigation to remember the following six points:

1. Examine facts as they are, not as you feel they should be or as they are said to be. Validate evidence in accordance with perception v. truth practice.

2. Avoid fitting the evidence to preconceived solutions and challenge everything.

3. Judgement must be part of every selection of alternatives, but do not use judgement when facts can replace it.

4. Look for ideas, hunches – try to innovate – but do not be unrealistic.

5. Involve others, particularly those who will have to authorise and operate the final solution. Analysis is a learning process.

6. The right questions produce good answers quickly.

The philosophy concerning the examination of problem areas is summarised in Fig. 4.11. The critical examination of what, where, when, who, how and why is applied to problem solving too and is also summarised in Fig. 4.11.

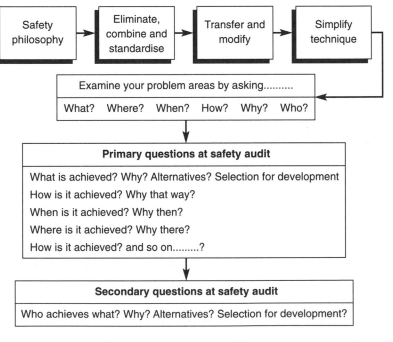

Figure 4.11 Safety audit philosophy

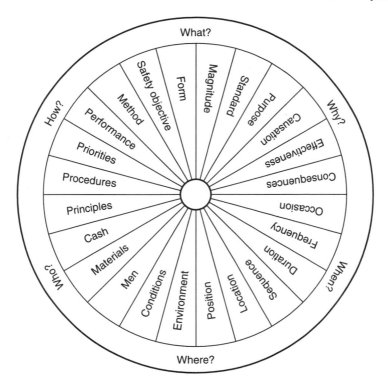

Figure 4.12 Mental roulette!

A standard critical examination sheet or 'roulette wheel' is shown in Fig. 4.12.

There are some common reasons why safety management solutions might be inadequate and it is only by studying how past solutions were arrived at and made that the thinking process can develop so that future results can be improved. Some of the most common categories into which unwanted outcomes can be placed are:

- Failure to identify true needs.
- Starting to develop before a satisfactory specification of needs is prepared.
- Inadequate knowledge of techniques, systems, people available, their costs, skill requirements, strengths and weaknesses and pay-offs.
- Lack of knowledge or awareness of basic principles and concepts.
- Lack of knowledge of other management functions.
- Lack of knowledge of alternative established approaches.
- Lack of time to make the best decision.
- Taking decisions in areas where one's own knowledge is insufficiently developed.

- Lack of knowledge of what is required of them in those implementing change.
- Inability to overcome personal prejudice.
- Failure to determine priority areas for effort and stick to them.
- Failure to make contacts or to use them.
- Lack of data search.
- Lack of adequate accident and other relevant data collection.
- Failure to communicate between those who are involved in the diagnosis and development of new approaches.
- Inability to profit from past mistakes.
- Lack of interest in increasing personal capability by adding to, questioning, rearranging one's beliefs and understanding of personal mental process.
- Failure to make maximum use for other purposes of every element of work done.
- Shortage of relevant data.
- Inadequate management policies, plans and controls.
- Inability to fix realistic targets and to check achievements.
- Striving for an inadequate concept of good solution.
- Allowing temporary expedients to become permanent.
- Poor human relations and motivation.
- Inadequate selection and training of staff.

CLIENT RELATIONSHIP IN PROBLEM SOLVING

Safety managers must sometimes bring in outside help when problem solving. It is also common for a specialist safety manager, say from head office, who arrives at a distant outstation to be known by name only. In such circumstances the safety manager then adopts the role of consultant as far as the outstation is concerned. Additionally, it is recognised that certain areas within the scope of the general safety practitioner may sometimes require expert help in a specific area outside the general terms of reference of the safety personnel available. Outside help will mean employing a consultant and there are several stages to consider when developing a consultancy relationship irrespective of whether the consultant is 'in-house' or from a completely external source. There are six basic stages to this which are:

1. Initial entry.
2. Contracting.
3. Diagnosis.
4. Action planning.

5. Evaluation.
6. Withdrawal.

Initial entry

It is essential that trust, confidence, credibility and integrity are established and there must be mutual feelings and understanding about the problem to be solved. This will necessitate the clarification and understanding of skills and experience being offered. Gaining mutual confidence that you are the person best suited to help is important. It is hoped that the desired outcomes are:

- Understanding of the problem.
- Mutual wish to work together.
- Beginning of a relationship.

Where an external consultant or internal safety manager is concerned it is crucial that their involvement is widely seen by the workforce as one employed purely to solve a particular safety problem and not for any other reason.

Contracting
Developing an initial contract on both a business and psychological basis is important. At this stage it is helpful to define helper/client roles, agree authority and responsibility aspects of the project and to agree the success criteria. From this stage it is possible to come to a formal contract, produce a plan of action, reaffirm commitment and the 'client' accepts the 'consultant' as the legitimate help.

Diagnosis
The 'consultant' plays the lead in determining who is the real client, what the specific problem(s) is/are and provides an opportunity to share new approaches, ideas and viewpoints on the problem. This is so that a joint definition of the problem can be published, ownership of the problem identified and agreement upon the evaluation criteria reached.

Action planning
This stage commences with a joint lead involving the safety manager and 'external' consultant in which the input from the latter diminishes as the action plans are implemented. This allows for the exploration of alternative methods and plans and for assessing consequences. At the same time leverage points can be identified before deciding upon methods and plans for the first action steps and the development of feedback systems for monitoring results. Outcomes from this stage should provide for a shared knowledge of all alternative strategies, a commitment to the selected plan and permit the weaning policy to be established (i.e. for a move away from dependency).

Evaluation

This will allow for the monitoring of people's reactions to plans as well as technical performance of safety programmes. At the same time, there should be assessment of resistance to change so that plan modification or success criteria can be achieved. In safety terms, this stage provides an opportunity to cultivate people's awareness of your concern and appreciation of their problems and to evaluate the level of dependency. This process will allow an opportunity for further action and also allow a reduction of people's worry levels. An openness and constructiveness in response is hoped for which is a basis for reducing 'consultant' dependency.

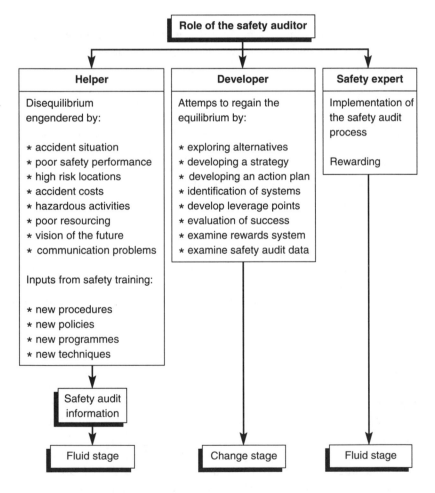

Figure 4.13 Change process in some organisations

Withdrawal

This stage will cater for the evaluation and the development of resources necessary for a 'no dependency' state to be achieved. At the same time, the client may come to an agreement concerning the planned withdrawal of the consultant whilst also assessing the organisation's self-confidence. It is hoped that such withdrawal will be without any side effects although there will be provision for re-entry where appropriate, especially for the continued successful operation of safety systems.

Changes in people can be classified under two headings. These are:

Gap bridging. Against an overall goal, a strategic need for change is established and the aim is to bridge the gap from where we are now to where we wish to be. The danger in many change situations arising during problem solving is that the 'gap' is seen as too large for those involved and the associated goals are refuted.

Incremental change. This is geared to reducing areas of unpleasantness, hurt, weakness but this process oscillates and is unrelated to an overall goal.

The change process in organisations is shown in Fig. 4.13.

Some notes summarising how to increase the motivation for change are given in Fig. 4.14.

Preparing for entry

When visiting an outstation as a 'consultant', or when the safety manager is undertaking duties in a consultative role, it is important to get the behaviour right. By thinking through some of the issues listed below, behaviour can be selected more appropriately. This may help to avoid some major pitfalls.

1. Get agreement on:
   - Objectives   – mutual understanding is important.
   - Who will be there   – do not be assailed without being ready.
   - Who should be seen – for what purpose?
   - How long will it last – people are busy, therefore keep
                              interviews as short as possible.

2. Think through:
   - Own objectives, needs and goals – what action is required?
   - Criteria for assessing help levels – commitments, skills, learning, payoffs for the organisation, for the consultant, for the department, etc.
   - Internal state at the moment – problems which are experienced from other internal/external people which may interfere with the meeting. Preparing to be non-defensive (exploring attacks and resistances rather than fighting back).

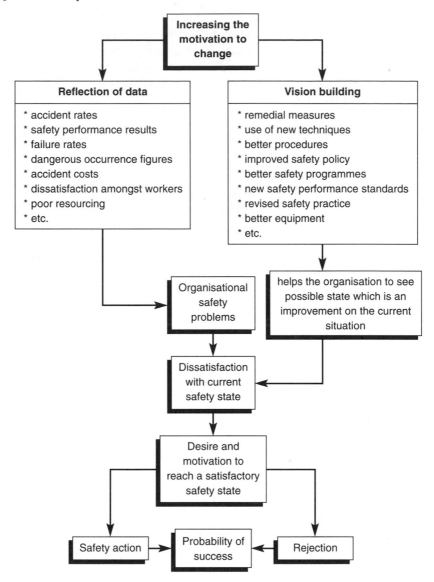

Figure 4.14 Increasing safety motivation

3. Try to predict:
    - Clients needs and goals – underlying feelings (e.g. use to which the consultant is being put).
    - Criteria for judgement – comparable experiences, appearance, sociability, self-confidence, publications, record of success.

Some common mistakes made by a safety consultant or outside safety expert can be summarised as follows:

- Losing professional detachment.
- Imposing own values.
- Becoming trapped in one part of the project.
- Unacceptable attachment to the client.
- Failing to seek help.
- Failing to recognise or analyse personal client resistance.
- Not being sufficiently candid.
- Not appreciating client's wish not to change.
- Changing only one subsystem.
- Unbalanced use of structural v. process change.
- Attempting an unbalanced change.
- Creating a change overload.
- Not recognising change resistance factors.
- Failing to recognise full implication accident analysis findings.

OPERATIONAL SAFETY MANAGEMENT ISSUES

The problem with operational safety management is developing a satisfactory conceptual framework which will enable organisations having different objectives but similar problems to be studied together. In theory, all operating systems can be regarded as being either input, transformation or output systems. This is sometimes called the conversion subsystem. The most common application of operational safety management is in manufacturing industries where the operating system is usually called the production system. However, the types of problems which occur in manufacturing industries also occur in service industries and organisations generally, and these can also be represented by the input, transformation and output model. Operating systems also have a controller feedback loop sometimes referred to as the controller subsystem. The feedback system consists of sensor, comparer, memory and effector functions. Operating systems are composed of smaller systems which may be connected in various ways such as in flow (series) systems or parallel systems.

Techniques are required to deal with problems of designing, planning and controlling safety operating systems. In practice there are three major reasons why manufacturing safety systems can be different from service safety systems:

1. Production is insulated from the environment;
2. Manufacturing safety systems do not usually involve the public;
3. The production function is usually more readily distinguishable.

How operations safety management is used for decision making is illustrated in the Handbook of Safety Management in Chapter 4.

DECISION MAKING IN OPERATION SAFETY MANAGEMENT

The problem lies in providing useful conceptual tools to aid the decision-making process. Making operational decisions is usually done in conditions of uncertainty or unreliability. Decisions are made in the context of the objectives of the organisation and these are both value and functional objectives and shape the criteria on which choice is made. The decision situation has five basic elements:

1. Strategies (alternative options).
2. States of nature (described by probability distributions).
3. Outcomes (for each combination of a strategy and a state of nature).
4. Forecasts (the probability of each state of nature).
5. Criteria (often conflicting).

The decision situation is formalised in the form of a matrix and for each strategy, all the various outcomes for possible states of nature are calculated and weighted by their probability of occurrence. In this way, the expected outcome for each strategy can be calculated. Also, the risk inherent in each strategy can be assessed together with the degree of difference between the expected outcomes of the different strategies. In practice, some states of nature do not apply to all strategies. There are usually more than one criterion which often are conflicting or non-quantifiable. The usefulness of the 'answers' depends on the accuracy of the forecasts.

Decision theory can provide valuable insights into a decision situation. It does not, however, make decisions on your behalf.

WORK MEASUREMENT

It is necessary to know how long a job will, or should, take so that manning levels can be calculated, labour costs can be estimated, machine loading can be done, safety can be assessed and payment schemes can be devised. Work measurement can be assed attempts to establish times for jobs under conditions of:

- Qualified workers.
- Specified jobs.
- A defined level of performance.
- Safety.

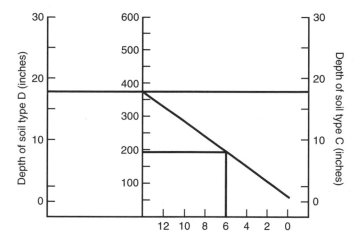

Source: Handbook of Safety Management, page 154.

Figure 4.15 A work measurement example

The technique involves basic times for jobs which can be obtained by any of the five methods employed in work measurement. These are:

1. Time study – using a stop watch.
2. Synthesis – from existing elemental data.
3. Predetermined motion – time systems.
4. Analytical estimating.
5. Activity sampling.

Standard times for jobs are obtained by adding relaxation allowances to the basic times. In practice, different techniques have different advantages depending on whether accuracy or consistency of times is required. The process of rating is still a very dubious procedure and it should be remembered that job timing can cause industrial disputes. Here the safety manager must ensure the fine balance between the need for his organisation to be as efficient as possible but within the maximum safety limits available. An example of work measurement for synthesising times for laying gas mains is given in Fig. 4.15.

Method study

The problem here is the development of an approach by which safer, more effective and lower cost methods of work may be devised. There are six steps to consider in method study. These are:

1. Select.
2. Record.
3. Examine.
4. Develop.
5. Install.
6. Maintain.

Whilst these are self-explanatory notes are attached below which may be added to these areas.

Select
This concerns the task or tasks to be studied. Monitoring of various tasks will highlight potential or substantiated danger.

Record
Note each existing method. This can be done in either of two ways:

By charts:
- Outline process chart.
- Flow process chart.
- Two-handed process chart.
- Multiple activity chart.

By diagrams:
- Layout diagrams.
- Flow diagrams.
- String diagrams.
- Memo-motion (time lapse).
- Micro-motion photography.
- Cyclographs.
- Cronocyclographs.

Examine
All methods should be examined together with obtained data and information. Use critical examination technique.

Develop
Develop new methods which will employ the principles of motion economy, safety and schemes for collecting ideas (brainstorming).

Install
Install new safe methods. Problems can be reduced by consultation with all concerned parties from the first stage when selecting jobs for study.

Maintain
As expected, it is necessary periodically to check and update methods.

Ergonomics

It is necessary to establish guide rules for the safe design of the physical environment and of equipment. Ergonomics is concerned with the physiological characteristics of people and how they interface with the equipment they use and the physical environment. The man-machine interface is concerned with information flows through displays and controls. The man-environment interface is concerned with how performance varies with environmental conditions. Techniques include information on:

- Effective designs of displays.
- The speed, accuracy, loading and range controls.
- The physical characteristics of people referred to as anthropometric data.
- Acceptable standards of heating, lighting and sound levels.
- General health and safety.

In practice, the redesign of equipment and working environment can be expensive. Designers find that designing for a range of physical sizes and individual preference can prove difficult.

JOB EVALUATION

This concerns the process of determining without regard to personalities, the worth an organisation places on one job in relation to that of another. The International Labour Office (ILO) states that job evaluation may be defined as an attempt to determine and compare the demands which the normal performance of particular jobs makes on normal workers without taking account of the individual abilities or performances of the workers concerned. There is a need to evaluate jobs because most people have strong feelings about 'fairness' or 'equity' of relative wage levels. Comparison of the contents of different jobs tends to influence attitudes. Workers doing the same job with equal efficiency under the same conditions should get the same wage and differences in the content of jobs should be reflected in the rates of pay for those jobs.

Job evaluation considers five fundamental issues.

1. A thorough examination of the job to be assessed.
2. The preparation of a job description to record its characteristics fairly.
3. The comparison of one job with another by an approved method.
4. The arranging of jobs in a progression.
5. Relating the progression of jobs to a money scale.

There are five common methods for job evaluation and these are:

- Ranking.
- Classification.
- Analytical methods.
- Factor comparison methods.
- Points rating method.

Safety should feature in all these methods but is sometimes overlooked by the job evaluator. Some safety audit processes use the point rating system therefore an example using this type of job evaluation method is given below.

Safety audit scoring example

1. Previous experience. First decide if any previous experience is necessary before the operator can perform the job in question. Unskilled or repetitive operations are not expected to require any previous experience and are catered for in item 2. below. When previous experience is necessary, decide what experience a new operator should have when applying for the job in the first place.

   Points range: 0–22.

2. Learning period. This is a measure of the time required for an individual to become familiar with new surroundings. Where no previous experience is required, it is necessary to consider the time that the 'average' untrained labour would need before the operator could produce enough to earn the minimum wage. Points may be awarded as follows:

   2 months 8 points
   3 months 10 points
   4 months 12 points

   Points range:
   0–12

3. Reasoning ability. How does the job require the operator to think? Is it a process that can be performed subconsciously or must the operator reason out a sequence of events?

   Points range:
   Low 0–4
   Medium 5–8
   High 9–16
   Very high 17–23

4. Complexity process. The extent to which the operation requires the mastery of an unusual number of details or the memorising of large numbers of variations.

Points range:
Low 0
Medium 0–3
High 4–7
Very high 8–11

5. Manual dexterity and motor accuracy. This is included to consider the unusual quickness or deftness which may be required to do the work successfully.

Points range:
Low 0
Medium 0–4
High 5–9
Very high 10–14

6. Materials. Measure the potential damage that can be caused to the materials or components of the product. This does not include tools or equipment.

Points range:
Under £100 0
Under £1,000 3
Over £1,000 6

7. Effect on subsequent operations. Consider whether lack of care might make work on subsequent operations more difficult or might necessitate additional operations.

Points range:
Low 0–2
Medium 3–5
High 6–8
Very high 9–12

8. Equipment. Refers to the damage that might be caused to machines or other items that assist in processing the product. It is necessary to value the damage that can be caused to the equipment by carelessness. As a useful guide and depending upon the type of industry involved, say 1 point per £1,000.

Point range:
0–x

9. Teamwork. Does the job demand close co-operation and adjustment of movement with other workers?

Points range:
2 in team 1
5 in team 2
10 or more 3

10. Attention to orders. To what extent does the job demand an ability to follow the details of a written order, a specification or the detail of a drawing?

Points range:
None 0
Occasional (simple) 1–4
Occasional (complex) 5–8
Constant (simple) 5–8
Constant (complex) 9–12

11. Monotony. This will consider whether the job is boring, tedious or irsome.

Points range:
0–7

12. Abnormal position. Are there any unusual or cramped positions required which cramps or strains muscles to an abnormal degree?

Points range:
Standing with little body movement 0–1
Frequent stooping or reaching 2–3
Continuous stooping, reaching/confinement 4–5

13. Abnormal effort. Is the work sufficiently heavy that the job is unattractive or that unusual physical strength is required. The actual weights to be lifted are not the important factor here but the physical effort needed.

Points range:
Lifting 20lb frequently or 50lb occasionally (low) 0–3
Lifting 50lb frequently or 100lb occasionally (medium) 4–7
Lifting 100lb frequently or 150lb occasionally (high) 8–10

14. Disagreeableness. This concerns the degree to which the job is made particularly unpleasant by the presence of wet, heat, cold, noise, dust and/or fumes.

Points range:
Low 0–2
Medium 3–5
High 6–8
Very high 9–10

15. Accident. Here the probability of risk must be assessed and then the potential severity of injury to the operator. It is recommended that the severity of injury is less important than the number of accidents occurring as the criteria between what makes a minor accident more serious are difficult to define.

To deal with this aspect equitably, company accident records must be consulted and a minimum of three years data should be used in calculating any trends. As a general guide the following pointing system might be considered:

Points range:

Less than 5 reported accidents/dangerous occurrences in past year 0–5
More than 5 but less than 10 6–10
More than 10 accidents (2 points per accident) 11–X

16. Disease. To what degree the working conditions are harmful to health or conducive to an occupational disease such as dermatitis, skin cancer, silicosis, poisoning by lead or aniline, etc. Again a retrospective study of company records is required to assist in weighting this fairly. The weighting can be split between potential v. actual. If a company uses known hazardous products but has no record of actual disease then standard weighting will apply. If disease is known to exist then the weighting may double.

Points range:

No recorded diseases 0–6
Recorded diseases 7–X

The assessment of manual jobs is usually carried out by a committee comprising of the departmental manager, job analyst, foreman and the workers' representative. Safety managers are not always consulted on these issues and where this happens the risk of accident potentiality being incorrectly assessed increases thus invalidating the job evaluation exercise. Methods for converting these points into money vary considerably and are discussed further in Chapter 7.

JOB DESCRIPTIONS AND SPECIFICATIONS

It is required that safety managers understand some of the human resource function within an organisation and how to use it effectively. Human resources management is difficult to define but generally comes in two parts. First, the theory, and second, the practice. This division has influenced the way in which the subject has been written about and more important, the way in which it is taught to students. What this means is that a subject such as motivation can be taught as a subject in itself without necessarily relating the theory to the management of people or relating it to, for example, the introduction of an appraisal system or a change in the way in which employees might be paid. A management decision to introduce such changes is influenced by our ideas on what actually motivates people to work. A safety manager should be able to apply his knowledge of people in the workplace if he is to manage the human resource effectively. People are rarely thought of in the same way as, say, equipment, time, cash, etc., possibly because people are flexible, different, creative and able to learn from their own behaviour. In addition, it is human resources which create, organise, control and co-ordinate all other resources in order to meet organisational aims and objectives.

Human resources form the basis of organisational activity, thus it is important for safety managers to learn and understand all they can about factors affecting people's behaviour at work.

Such disciplines involve the study of psychology, sociology, economics, politics and anthropology. These disciplines combine to provide a collection of findings, ideas, methods and approaches to the development of knowledge as to why people behave as they do and an awareness of the implications this has for management. It is important to remember that an aspect of theory may affect one or more practical situations or that a practical problem may draw upon many theoretical inputs. Although motivation is followed by remuneration and appraisal, questions of manpower, planning, organisational structure and training also draw upon motivational theory.

The study of systems theory has had a major impact on the way people have approached the study of organisational behaviour. In the beginning, the study of organisational behaviour centred on individual behaviour but as research developed, psychologists and sociologists came to realise that an organisation was a complex social system which required study as a total system if individual behaviour within it was to be understood. It is a traditional human failing to assume that any event (like the accident) stems from a single cause. There is always a tendency to simplify cause and effect relationships. This can be illustrated by the development of scientific management approaches to the installation of new technology which tend to focus on ultimate efficiency at the expense of the human and social aspects of the work situation. From this, human relations theory stressed that informal work groups are important as is the need for management to communicate with the workforce. This theory neglected the technological aspects of organisations generally and both approaches failed to consider the effects of the environment and cultural norms upon organisational behaviour. In order to assist the safety manager to operate more effectively, there is a need to look at what determines the behaviour and responses of individuals within organisation, and how they can become more effective in achieving their aims and objectives. In order to do this it is necessary to move away from thinking that events have single causes and approach organisational behaviour from the concept of systems theory. This has as its basis the notion that all events are determined by multiple contributory factors (like the accident) and are interdependent.

Organisations are constantly involved in the process of attempting to select and recruit appropriate people for the right job. This process has become sophisticated and involves a considerable amount of time and effort particularly if effective recruitment and selection techniques are to be developed. A basic outline of the recruitment and selection process is discussed below but it must be stressed that each organisation will need to develop the process for its own ends. However, the fundamentals remain the same and the basic stages to consider in the recruitment and selection process are:

- Job analysis or description: have an understanding of the job to be filled.
- Job specification: understand the knowledge, skills and aptitudes required to do the job.
- Recruitment media: decide where suitable applicants are to be found.
- Advertising: decide how people can be persuaded to apply.
- Job administration: choose the means of finding whether the applicants have the required knowledge, skills and aptitudes. Consider methods of application such as application forms, interviews, references, selection tests, etc.
- Rating schemes: make a choice of which method to be used for scoring applicants.
- Induction: introduce the successful candidate to the job and to the organisation.
- Evaluation: assess the recruitment and selection process.

The features of the recruitment and selection process outlined above are now discussed in more detail.

Job description

This process examines a job in order to establish its component parts and identify the circumstances in which it is performed. It should also answer the very basic question as to whether the job is really needed or not. Methods used for carrying out job analysis are as follows:

- Via a process of interviews or group discussion information should be gathered to ascertain what the job holder states that his job is.

- Via a similar process ascertain what the job holder's manager thinks the job is.

- Using activity diaries, activity sampling techniques, films, critical incidents and/or exit interviews find out what the job holder's job actually is.

- Conduct interviews with both the job holder, manager and independent adviser (where appropriate), using company information to decide what the job should be.

It is important to develop a list of actual duties and activities rather than amass a list of general responsibilities such as seeing a statement like 'he is responsible for several men'. The job description will result from the job analysis and it usually includes a broad outline statement of the purpose and scope of the job, together with a more detailed list of the activities and responsibilities involved.

Job specification

The job specification results from identifying the skills, knowledge and aptitudes required by a person to carry out their duties as described in the job description. It is a detailed statement of the physical and mental abilities required for the job and when relevant, social and physical environmental aspects of the job. When developing a job specification, it is important to remember that it is used as the basis of assessing a candidate, and should therefore be developed in such a way that the skills and knowledge lend themselves to being measured. It is an attempt to identify those human characteristics which are directly related to important aspects of job performance. The job specification is usually based on Rodger's 7-point plan or Munro Frazer's 5-point plan. Whichever plan a particular organisation uses it will normally consist of the following general headings:

- *Attainment:* School, further education, professional qualifications and experience.
- *Physical:* Health, appearance, manner, eyesight, fitness, etc.
- *Intelligence:* Judgement, problem solving, IQ, etc.
- *Special aptitudes:* Manual, written, verbal, language, etc.
- *Interests:* Temperament, adjustment, etc.
- *Circumstances:* Domestic, travelling, etc.

Not all the headings may be relevant to any one job, as a study of the job will usually decide on the importance and relevance of each feature. For example, fitness and physical strength is more important than educational qualifications for the employment of unskilled labouring jobs. It is also important to have some system of rating individuals against the criteria which are considered necessary so that comparisons can be drawn.

Job performance standards

Job analysis has a duty to consider the performance standards associated with the job. These will include safety performance standards. These standards serve two functions:

1. They are targets for employee efforts and the challenge of meeting objectives may serve as the motivation employees need. Once standards are met, employees feel a sense of accomplishment and achievement and the outcome contributes to worker satisfaction. Without standards employee performance will suffer.

2. Standards are criteria against which job success is measured. They are indispensable to safety managers who attempt to control the safety of work performance. Without these standards, no control system can evaluate job performance.

All such control systems have four features:

- standards;
- measures;
- correction;
- feedback.

The relationship between these four features is illustrated in Fig. 4.16. Job performance standards are developed from job analysis information and then actual worker performance is measured.

When measured performance varies from the job standard then managers must take corrective action immediately. This action serves as feedback to the standards and actual safety performance. This feedback leads to changes in either the standards themselves or the actual job performance and these standards are key to the control system.

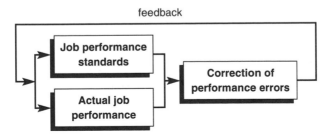

Figure 4.16 A job control system

Recruitment

This is the process by which organisations attract people to apply for job vacancies. The objective of any recruitment activity should be to attract a compact field of suitable candidates and to achieve an optimum balance between coverage and cost. There are various sources of potential candidates and these need to be considered if a sensible recruitment policy is to be developed.

Some problems associated with recruitment are:

- Choosing the right media.
- Obtaining the optimum number of candidates.
- The wording of the advertisement.
- The need to take account of the company image.
- The need to attract candidates and impel them to apply.

The advertisement

The first priority is to decide where to place the advertisement. The correct selection of media is a sophisticated business and is, therefore, often contracted out by organisations to outside agencies. The job advertisement has four roles. These are:

1. To attract a compact field of applicants who are capable of doing the job.
2. To inform suitable candidates about the job, the organisation and what is required.
3. To impress suitable candidates.
4. To impel candidates to apply.

Most job adverts feature the following information:

- Title of vacant post.
- Grade or salary.
- Name of the organisation.
- Details of the job.
- The type of person required.
- Information about the organisation.
- Benefits.
- Action to be taken to apply.
- Closing dates for applications.

It is now normal to see companies making statements as to company safety policy within some job adverts.

CRITICAL PATH ANALYSIS

This is a method of helping the safety manager understand the various phases to a task and why some tasks have to be undertaken before others. Critical path analysis (CPA) makes a valuable contribution to safety for it provides for things to happen in a logical order. As an example, a company is to modify its works' first-aid station and install 'supernurse', a new, all-purpose, automatic bandaging and plastering machine. The general manager shows you a list of jobs to be done and seeks your comments as to the safety aspects of installation. Having a copy of the list you also consider the most efficient way of doing the job whilst at the same time as maximising safety. The list is as follows:

Job	Description	Duration
a	order 'supernurse'	20 days
b	clear first aid room!	1 day
c	reinforce the floor	4 days
d	install nurse controller	2 days
e	install electrics	5 days
f	install overhead bandage feed	3 days
g	install unfortunate casualty	2 days
h	set and adjust controls	3 days
i	fit autoplaster pump	1 day
j	fit autoplaster spreader	1 day
k	connect autoplaster pump/spreader	1 day
l	test on patient and adjustment	3 days

In such a case it is important to be aware of the relationship between all of the listed activities. For example, activity l (casualty testing) cannot occur until task g (install unfortunate casualty) has been completed. Activity k cannot take place until both j and i have been satisfactorily completed. It may be dangerous to have the electrician laying wires down when men are working off ground fitting the overhead bandage feed so (e) has been placed in front of (f). For each activity then we have:

Activity	Preceding activities
a	–
b	–
c	b
d	e
e	c
f	c
g	a,f,d
h	g
i	e
j	g
k	j,i
l	h,k

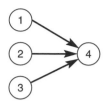

A single task cannot be achieved until
all the activities leading up to it are completed
safely. Safety audits need to understand that
event 4 cannot be reached until activities 1-4
2-4 and 3-4 are all safely completed.

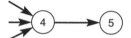

No activity can commence until the tail event
is reached. Activity 4-5 cannot start until event
4 is reached. Two different conventions are
that time flows from left to right and head events
have a number higher than the tail.

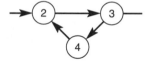

This event is impossible since event 2 depends on
event 3 which in turn depends on event 4 and this
relies on event 2 and so on. This is referred to as
looping. Safety audits need to be able to identify
these loops which can be difficult to identify in
more complex situations.

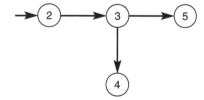

This is referred to as 'dangling' and can be identified
as a logical fault since activity 3-4 leads to nothing.

Figure 4.17 Some timepaths through a network for safety audit purposes

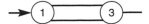

These are referred to as 'dummies' which
are activities which neither use resources
nor time and are used on two occasions. These
are:

1. When two independent activities have the
same head and tail as shown above. In order
to reduce confusion, a dummy is introduced
as follows:

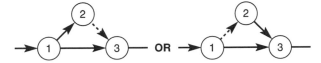

2. When two independent chains have a common
event as shown below.

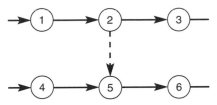

Figure 4.17 cont. Some timepaths through a network for safety audit purposes

A useful way of illustrating the relationships given above is by drawing a
network of arrows where each arrow indicates an activity. These have been
illustrated in Chapter 4 of the Handbook of Safety Management and these
processes should be understood by the safety auditor in order to identify
appropriate critical paths. More commonly, arrows are used to show activities whilst circles are used to show events.

Numbers shown by the arrows indicate the length of time or duration that
the activity takes. The whole task described above can be displayed diagrammatically which makes analysis much simpler to understand. When
drawing networks, there are some conventions in general use:

- Networks are composed of events and activities and an event is summarised as a definite recognisable situation; a point in time. One event
  can mark the concurrence of several separate events and can therefore
  sometimes be referred to as a node or junction. However, an event cannot
  be reached until all activities leading to it are complete.

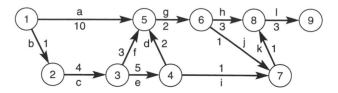

The 'supernurse' example illustrates the shortest time in which the total task can be completed is 20 days. This is the length of the longest path through the network being the sequence of safety audit events i.e. 1-2-3-4-5-6-7-8-9 or the activities b,c,e,d,g,h, and I. This is called the critical path.

Figure 4.18 Identifying the critical path for safety audit purposes

- An activity is the work or task which leads to the event. On occasions an activity is not true work since no resources are consumed. For example, waiting for the delivery of the 'supernurse'. This is not a true job but is included in the network because the activity is vital to the project. Events which occur at the beginning and end of an activity are referred to as the tail and head events.

Analysing the network

A network illustrates what is to be done, when, by whom and in what time. It also provides a clear statement of policy which can be readily understood by potential users. If no further action were taken, considerable benefits would have already been derived. By using simple arithmetic, it is possible to extract a considerable amount of information. The critical path is the sequence of events which indicate the shortest time in which the total task can be completed (i.e. the longest timepath through the network). See Figs. 4.17 A to D.

The importance of the critical path comes from the fact that only on this sequence of activities will any delay to an individual activity necessarily lead to an extension in the total time set aside for the project (see Fig. 4.18). All other activities have some degree of flexibility associated with them and this can be quantified by calculating the earliest time that an activity can begin.

Fig. 4.19 shows a critical path from 0 to 1 to 2 to 4 to 5, giving a total project time of 14 days. By dividing up the circles shown, it is possible to consider the earliest time an event can occur. The earliest event time (EET) of event 0 is 0. Since activity 0–1 takes 4 days, the EET of event 1 will be 4. Similarly, the EET's of events 2 and 3 are day 9 and day 3 respectively. There

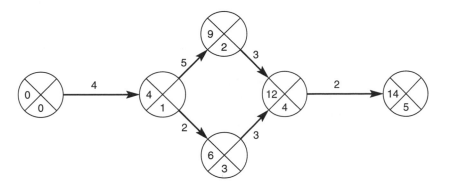

Figure 4.19 The float

are two routes to event 4 since an event cannot occur until all preceding activities are completed, the EET for event 4 must be derived from the EET for event 2, i.e. 9+3=12.

If this process is reversed, it is possible to find the latest time an event can occur. Event 4 must not occur after day 12 if event 5 is not to be delayed and so on. These latest event times (LET's) are written in the right hand side of the circles as shown in Fig. 4.20. By examining activity 1-2 which is of 5 days duration it can be seen that it cannot start before day 4 and it must not finish after day 9. This leaves 5 days in which to complete a 5 day task. This is referred to as having no float. By examining activity 1-3 it can be seen that this job cannot start before day 4 and must finish before day 9. There are 5 days in which to complete a 2 day job therefore this activity has a 5-2=3 days float.

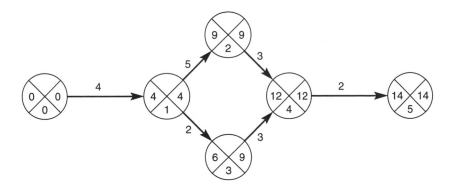

Figure 4.20

| 4 | 5 | 6 | 7 | 8 | 9 |

1 - 3

All the events on the critical path have the same earliest and latest event times. This illustrates that all the activities on the critical path have zero float.

Figure 4.21 No float?

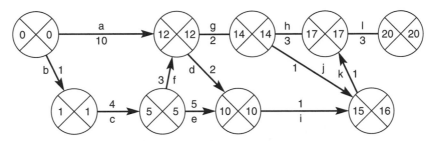

By following the same procedure for the original example, it can be calculated that the earliest and latest event timing and the float associated with each activity are as follows:

Activity	Start time		Finish time		Duration	Float
	Earliest	Latest	Earliest	Latest		
a	0	2	10	12	10	2
b	0	0	1	1	1	0
c	1	1	5	5	4	0
d	10	10	12	12	2	0
e	5	5	10	10	5	0
f	5	9	8	12	3	4
g	12	12	14	14	2	0
h	14	14	17	17	3	0
i	10	15	11	16	1	5
j	14	15	15	16	1	1
k	15	16	16	17	1	1
l	17	17	20	20	3	0

When an activity has a float, its start can be adjusted to suit other factors such as safety requirements or resource allocation. A float is more easily recognised if the network is illustrated as a *GANTT* chart. The 'supernurse' example is given in figure 4.23 as a *GANTT* chart.

Figure 4.22 The critical path

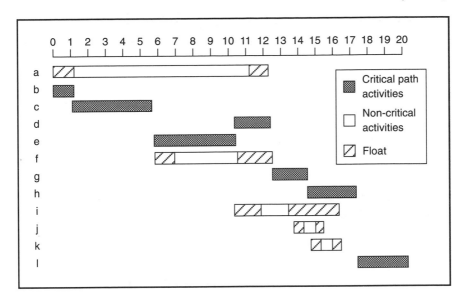

Figure 4.23 The 'supernurse' GANTT chart

SUMMARY

In this chapter an attempt has been made to outline the importance of operations management within the decision-making process and to highlight its importance to safety and the factors to be considered when carrying out a safety audit. Many organisations use operations management as a means of maximising efficiency but a danger exists that this could be at the expense of safety with the result of significantly downgrading the effectiveness of the decision.

Those responsible for safety within the workplace must have an input to all operations management studies so that the safety factor can be included within the overall plan. Connected with operations management programmes are matters relating to incentive schemes and other types of payment for reward.

5 AUDITING SAFETY POLICY

In this chapter we will look at the following:

Part 1: Is the Safety Policy towards resourcing correct?

- Legal requirements
- Statement of objectives
- Budgetary provision
- Staff
- Organisational structure
- Control of Substances Hazardous to Health (COSHH)
- First aid
- Heating
- Lighting
- Ventilation
- Noise and vibration

Part 2 Is the Safety Policy adequate?

- Decision-making environment
- Safety committees
- Management involvement
- Trade union liaison
- Liaison with other interested groups
- Industrial tribunals
- Disasters and emergency plans
- Statements on health and safety policy

PART 1

THE SAFETY AUDIT OF ORGANISATIONAL POLICY

An organisation's policy towards health and safety should be regularly reviewed so that it is certain that it conforms to the requirements of the law and satisfies the organisation's statement on health and safety policy. The methodology employed in conducting the safety policy review, however, is known to vary widely from rather sketchy attempts to very detailed pro-grammes. The emphasis on any audit, review or survey should be the time taken to carry it out. Some of the more detailed commercial safety auditing packages suffer from this aspect and require a large amount of staff time to complete properly. In the previous chapter it was suggested that the safety auditor needs to understand the individual task activities in order to conduct a successful safety audit. However it is not always possible, in some indus-trial contexts to do this but the safety manager should ensure that the infor-mation required is obtained from those departments within the organisation and that the information meets the criteria which have been discussed so far. The following safety policy areas are provided merely as suggestions for this aspect of the safety auditing process and should not be regarded as an exhaustive list. On the contrary, industries vary and the purpose of this chap-ter is to encourage the safety manager to develop a safety auditing package which will suit his own particular needs. Some of the main areas that need to be considered in the safety policy review are now discussed.

LEGAL REQUIREMENTS

As laws and regulations change systems within the organisation should also be flexible enough to change. New laws, regulations and other legal require-ments do not suddenly *appear* and sufficient time is usually given to allow for organisational policy changes to be made within an approved or agreed time scale. Some legal requirements are more relevant to some activities than others. It is a good policy to list all those regulations and other legal requirements that affect organisational policy in respect of health and safety and then to check that these documents are:

- held in a place of common access;
- available to be seen and read by the entire workforce.

New regulations or requirements *in the pipe-line* should also be available and appropriate warning orders issued.

STATEMENT OF OBJECTIVES

All organisations should publish a statement of objectives in terms of health and safety. In a modern society these need to reflect the needs of the industry at the time. A statement of objectives made in the 1970s may not necessarily be relevant for the 1990s particularly with advances in technology and the impact of any EEC requirements after 1992. A statement of objectives should clearly state:

- the organisational mission;
- its strategic approach to accident prevention and reduction;
- the tactics to be employed in accident prevention and reduction;
- the methods it will use to reduce or prevent accidents from happening in an operational setting.

The organisational mission

This is a simple statement of intent and should be clearly understood by everyone in the organisation. For example, the organisational mission might be:

'TO PREVENT ACCIDENTS FROM HAPPENING'

This statement is clear and easily understood. Statements should avoid making forecasts such as:

'TO REDUCE ACCIDENTS BY 2 PER CENT'

Such statements might look good from a PR or 'political' point of view but forecasts are usually highly inaccurate and in the final analysis arguments tend to surround the method of forecasting rather than the accident reduction objective.

Strategic objectives

This too is a simple statement listing the strategies to be used by the organisation in its attempts to reduce or prevent accidents from occurring. For example an organisation may wish to:

- develop a company policy concerning safety education, training and publicity;
- develop a company policy to prioritise areas for remedial accident prevention and reduction strategies in the organisation;
- implement a policy of budgetary provision for strategically identified safety problem areas;
- develop a safety policy.........and so on.

These need to be identified early on in the safety audit process and may be found in the company statement of safety policy. They must also be examined at a tactical and operational level within the company and a check made on validity and whether they are up to date.

BUDGETARY PROVISION

The safety budget should be sufficient to carry out the organisation's statement of objectives in relation to health and safety. It is necessary to examine the policy on budget preparation and to seek ways in which priorities are identified for accident prevention and reduction activities. Areas where shortfalls in budgetary resourcing can be identified must be recorded so that management are then in a position to review their policy in the future.

An integral part of the safety audit on the budget will include over/under spend and will note the activities on which resources have been allocated. It is important to compare the areas of safety activity with the accident situation in the organisation. It would not seem appropriate, for example, to spend 30 per cent of a safety budget on an activity occupying only a 2 per cent accident incident rate. However, if there are higher areas of accident incident rates it is important to compare budgetary allocation to these areas of activity.

STAFF

This part of the safety audit will examine individual and group (or team) performance standards. It is also important that at this stage, individual and group objectives are also examined. This will include the way in which performance standards and objectives are set and the way in which they are monitored and reported back to senior management.

Within this part of the safety auditing process, it is usual to examine the way in which management monitor and evaluate both individual and group performance standards and objectives. It is also necessary to examine welfare facilities and some notes on this follow.

Within the general terms of welfare it is necessary to provide sufficient and hygienic toilet and washing facilities in all places of work. There are additional requirements to provide washing facilities where the work is particularly dirty or arduous. These requirements will be found under the Factories Act, 1961, the Offices, Shops and Railway Premises Act, 1963 and the statutory instruments made under them. In addition, the HASAWA also requires that a safe and healthy working environment be provided and that adequate facilities and arrangements must be made for employees' welfare. This will include the provision of adequate and hygienic toilet and washing facilities.

All sanitary conveniences provided by an employer must be kept clean and there must be provision for lighting, ventilation and privacy. Where persons are employed of different sex then proper and separate accommodation

Minimum number of conveniences		How many at audit?
	Toilets	
For every 25 female employees	1	
For every 25 male employees	1	
Where the number of males exceeds 100 and sufficient urinal facilities are also provided	4	
For every 40 employees over 100	1	
Where the number of males exceeds 500 then for every 60	1	
Satisfactory/Unsatisfactory	Yes/No	

In counting the number of employees, any odd
numbers less than 25 or 40 are regarded as 25 or 40.

Figure 5.1 A summary of the Sanitary Accomodation Regulations, 1938 (amended) 1974, for safety audit purposes

must be provided. There are rules concerning the number of conveniences outlined in the Sanitary Accommodation Regulations, 1938 as amended in 1974. They are summarised in Fig. 5.1.

There is an obligation to provide and maintain proper washing facilities including the supply of hot and cold running water. In addition soap and clean towels (or hot air dryers) are also a requirement and must be kept in a clean and accessible condition (see s.58(1) of the Factories Act, 1961).

ORGANISATIONAL STRUCTURE

Is the organisational structure capable of meeting the performance standards and/or objectives which have been set? This is an important question to be considered by the safety auditor. This part of the safety audit would also examine the efficiency and effectiveness of hierarchical structures in order to test that decisions are able to be taken when needed. The exercise would normally include departmental structures, sections, groups, teams and individuals. Accident research has found that organisational structures can contribute to accidents, particularly in large industries and it is essential that such hier-archical structures come under close scrutiny by the safety audit team. This aspect of the audit would examine roles and tasks as discussed earlier in Chapter 3 and would also consider those issues raised in Chapter 4.

CONTROL OF SUBSTANCES HAZARDOUS TO HEALTH (COSHH)

The COSHH Regulations introduce a new legal framework for the control of substances hazardous to health in all types of businesses, including factories, farms, quarries, leisure and service activities, offices and shops. The Regula-tions require you to make an assessment of all work which is liable to expose any employee to hazardous solids, liquids, dusts, fumes, vapours, gases or micro-organisms. Assessment means evaluating the risks to health and then deciding on the action needed to remove or reduce those risks.

The responsibility to make the assessment rests with you – the employer. As the employer, you could do or lead the assessment yourself or give the task to someone else with the authority and ability to get all the necessary information and make the correct decisions about the risks present and the necessary precautions. That person should know the point of the various requirements of the COSHH Regulations and have access to a copy of the Regulations and approved code of practice. Whoever does the assessment, you should make sure that managers, supervisors and employees' safety rep-resentatives are fully consulted about the work processes, about what work-ers are doing (or are liable to be doing), and about the risks and the neces-sary precautions.

In some cases, particularly if you are in doubt over the answers to the fol-lowing questions, you may need to consult your supplier or trade association or even obtain expert advice about what substances are involved in the work. Ask yourself if employees are liable to be exposed to hazardous sub-stances in your workplace. Include service activities as well as production processes.

How can substances hazardous to health be identified?

- For substances brought in, check the safety information on the labels and the information for safe use provided by your suppliers (they are required by law to do so).
- Use your existing knowledge (e.g. past experience, knowledge of the process, understanding of current best practice in your industry, informa-tion on work-related health problems in your industry.
- Ask your trade association and other employers in the same business for their experience and advice.

Question	Corrosives	Acids	Solvents
What is brought into the workplace? What is used, worked or stored? What substances are produced at the end of the work process?			

	Dust	Fumes	Gases	Residues
What is given off during any process or work activity?				

	YES	NO
Has the Health and Safety Commission approved an occupational exposure standard for the substances and is it listed in Guidance Note EH40/89?		
Is it listed in Part 1A of 'Information Approved for the Classification, Packaging and Labelling of Dangerous Substances' as being very toxic, toxic, harmful, corrosive or irritant?		
Is the substance present at a substantial concentration in the air?		
Is there a minimum exposure limit to the substance?		
Is it a micro-organism which can cause illness?		

NB If the answer is 'Yes' to any of these questions, it may be a substance hazardous to health within the meaning of the COSHH Regulations. It must depend on whether it arises out of, or in connection with, work under the employer's control.

Figure 5.2 A typical COSHH preliminary checklist

- Check COSHH: is the substance mentioned in any of the Regulations or Schedules? Is it listed in HSE Guidance Note EH40?
- Examine published documentation, trade data, HSE guidance material.
- Check Part IA1 of the approved list issued under the Classification, Packaging and Labelling of Dangerous Substances Regulations 1984: anything listed as very toxic, toxic, corrosive, harmful or irritant comes under COSHH.

Do the ways in which each substance is handled or is present in the workplace give rise to any risks to health in practice now or in the future?

Observe, find out about and consider:

- Where and in what circumstances substances are used, handled, generated, released, etc. What happens to them in use?
- Is their form changed (e.g. solids reduced to dust by machining)? Identify locations (e.g. handling departments, storage areas, transport).

- What people are doing; what might they do?
- What measures are currently taken to control exposure and to check on the effectiveness and use of those measures?
- Who will be affected (e.g. employees, employers, contractors, public)?
- Is exposure liable to occur?
- Is it likely some of the substance will be breathed in?
- Is it likely to be swallowed following contamination of fingers and/or clothing?
- Is it likely to cause skin contamination or be absorbed through the skin?
- Is it reasonably foreseeable that an accidental leakage, spill or discharge could occur (e.g. through breakdowns of the plant or control measures or operators' mistakes)?

Reach conclusions about people's exposure: who, under what circumstances, the length of time they are or could be exposed for, the amount they are exposed to, and how likely exposure is to occur. Combine this with knowledge about the potential of the substance for causing harm (i.e. its hazard) to reach conclusions about the risks from exposure.

Sometimes, of course, the quantities, the exposure time or the effects are such that the substances do not or could not constitute a risk – but you must have the information to back up this conclusion.

Action to be taken

If the assessment shows that there is no likelihood of a risk to health, the assessment is complete and no further precautions are needed.

If the assessment shows that further action is needed, you have to decide what needs to be done to complete the assessment requirements. If it is reasonably practicable to do so, you should prevent anyone from being exposed to any hazardous substances. Where it is not reasonably practicable to prevent people being exposed, you have to ensure their exposure is adequately controlled and their health protected. In such cases you will need to:

- Select the measures to achieve and sustain adequate control.
- Work out arrangements to make sure those control measures are properly used and maintained.
- Make sure your workforce is trained and instructed in the risks and the precautions to take, so that they can work safely. In some circumstances, employees need to be monitored and arrangements made for them to be under health surveillance (check COSHH, HSE guidance notes relevant to your work and trade literature).

Unless you can easily report and explain your conclusions at any time because the assessment is simple and obvious, you should make a record of it. Record or attach sufficient information to show why decisions about risks and precautions have been arrived at and to make it clear to employees, engineers, managers, etc. what parts they have to play in the precautions.

If the conclusions alter, for example, the introduction of a new process or machine or a change in the substances used, or if there is any reason to suspect that the assessment is no longer correct, for example, reports of ill health related to work activities, the assessment must be reviewed to take account of these new circumstances.

FIRST AID

The obligation to provide first-aid facilities comes within the general duties under the Health and Safety at Work Act, 1974. This requires employers to ensure a healthy and safe working environment for their employees. Section 2(1) requires that employers will ensure, so far as is reasonably practicable, the health, safety and welfare of all their employees. This extends to the provision of appropriate first-aid facilities. The Health and Safety (First-Aid) Regulations, 1981 provide for the following:

- Provision of first-aid equipment and facilities.
- First-aiders.
- First-aid room where required.
- First-aid training.

First-aid equipment and facilities

An employer must provide adequate equipment and facilities to enable first-aid to be rendered to employees who:

- are injured at work; or
- become ill at work.

There are four criteria used to decide what provision is necessary:

1. The number of employees.
2. The nature of the business.
3. The size of the organisation and spread of employees.
4. Location of the business and employees' place of work.

A summary of the criteria is given in Fig. 5.3.

First aiders	Employees	How many at audit?
1 2	50 150	
First-aid room	Employees	
1	250	

Satisfactory/Unsatisfactory?	Yes/No

Where premises are said to be hazardous such as shipbuilding, factories, warehouses and farms there should be a trained first-aider as outlined above at all times. Where a first-aider is away or off work, cover for the absence must be provided by the employer. Where an organisation employs fewer than 50 persons then an employer must still provide for first aid cover.

Figure 5.3 First-aid cover requirements

First aiders

The first aid regulations require that an employer must provide, or make available, an adequate number of suitable people who have been trained in first aid. A person is not regarded as suitable unless that person has undertaken a course of first-aid training which has:

- been approved by the Health and Safety Executive; or
- been provided with additional specialised training so approved.

A suitable person may be:

- a first aider;
- an occupational first aider;
- another person who has approved training and qualifications (e.g. Registered General Nurse or medical practitioner).

A first aider is a person who has received approved training and holds a valid first-aid certificate.

First-aid rooms

Such a facility is required where 400 or more employees are at work. It is expected that such a facility will be appropriately staffed. Appropriate staff requires a minimum qualification of an occupational first aider where:

- there are establishments with special hazards;
- the place of work is a construction site with more than 250 persons at work;
- access to casualty centres is difficult.

Otherwise, a certified first aider is necessary.

Access to a first-aid room should be available at all times when at work. Such a room should be sited in an appropriate location so as to facilitate vehicular access so that injured persons may be transported to hospital with the minimum of inconvenience to everyone, including the workforce. Such a room should:

- contain suitable first-aid equipment and facilities;
- be properly ventilated;
- be adequately heated;
- have sufficient lighting;
- be clean and tidy;
- be properly maintained;
- be large enough to house a couch;
- have suitable access and egress for a stretcher, wheelchair or carrying chair;
- indicate clearly the names and locations of the nearest first aiders.

First-aid boxes should be the responsibility of the qualified first aider. Do not leave first-aid boxes open and unchecked otherwise when they come to be used they will probably be empty!

Self-employed personnel are required to provide themselves with adequate first-aid facilities and as such should carry their own first-aid kit. Professional drivers who are employees or self-employed should also be provided with/or provide for themselves (if self-employed) an approved first-aid kit for use in their vehicle whilst away from their depot or works.

First-aid training

It is essential that there are sufficient qualified first aiders available on duty at any one time to allow for absences. Unless an organisation has sufficient numbers to train, it would probably not be financially viable to run an in-house course. However, if such a localised demand exists, then the course content, trainers and testing facilities will need to be approved by the Health and Safety Executive. It is usual for the smaller company to contact:

- the St John's Ambulance Brigade;
- the British Red Cross;
- the Health Promotion Manager of your local Health Authority.

All of these organisations will be able to advise on appropriate approved certification first-aid courses.

HEATING

The legal requirements covering the provisions in respect of temperature in places of work is found in section 3 of the Factories Act, 1961 and in the Offices, Shops and Railway Premises Act, 1963. Also, within the general terms of the Health and Safety at Work Act, 1974 every employer is required to provide and maintain a safe and healthy working environment. This includes temperature and humidity. A thermometer must be provided and the minimum acceptable temperature in an environment where workers are expected to sit for a great proportion of their time is 16°C. This temperature must be reached and maintained within one hour from the first hour of work.

There are specific regulations and statutory obligations relating to different industries such as the textile industry or where artificial humidity is produced. A general summary is provided in Fig. 5.4.

LIGHTING

Most modern buildings will meet the legal requirements regarding the provision of adequate lighting. The requirements to provide good and adequate lighting can be found in the Factories Act, 1961, the Offices, Shops and

Workplace	Temperature	Essential requirements	Temperature at audit?
Office	16°C (min)	Thermometer	
Factory	16°C (min)	Thermometer	
Steam room	22.5°C (max)	Hygrometer	
Other room where artificial humidity is produced	22.5°C (max)	Hygrometer	

Satisfactory/Unsatisfactory?	Yes/No

Figure 5.4 Some common temperature requirements

Type of lamp	Lumens per watt	Location	Number	Date last checked
Incandescent	10 to 18			
Tungsten halogen	22			
High pressure mercury	25 to 55			
Tubular fluorescent	30 to 80			
Mercury halide	60 to 80			
High pressure sodium	100			

Satisfactory/Unsatisfactory?	Yes/No

Incandescent lamps are the common coiled filament lamps.
N.B. At safety audit these sheets need to be checked

Figure 5.5 A standard lighting checksheet

Railway Premises Act, 1963 with further requirements to provide a healthy and safe working environment within the terms of the Health and Safety at Work Act, 1974. If an employer fails to provide and maintain a suitable standard of lighting he will be in breach of Section 5(1) of the Factories Act, 1961. It is important that personnel managers set up a system whereby lighting (e.g. bulb replacement) is monitored regularly and that light bulbs provide sufficient lumens per watt where natural lighting is deficient. It is not the intention of this section of the book to produce a lighting engineer but to provide sufficient information to monitor information. To this end Fig. 5.5 summarises all that the manager need know for daily operational purposes.

Further advice, where necessary, should be sought from an appropriately qualified lighting engineer at safety audit.

VENTILATION

Section 2(1) of the Health and Safety at Work Act, 1974 is regarded to include a duty to provide all workers with an adequate supply of pollution or uncontaminated free air. Because employers must provide and maintain a safe and healthy working environment, they may be criminally liable if the workplace is not adequately ventilated and dust or contaminant free. Moreover, in section 2(2) of the HASAWA all employers must inform, instruct and train their staff in health and safety procedures. This means that employees must know how to use, test and maintain equipment that ensures a pure air supply and controls air pollutants such as dust and fumes. In addition, Sections 4 and 63 of the Factories Act, 1961 contain two separate provisions in respect of:

- Air purification (Section 4);
- Dust and fumes (Section 63).

Air purification

A circulation of fresh air is necessary to:

- ensure and maintain the adequate ventilation of places of work;
- render harmless, as far as possible, all fumes, dust and other impurities which may be harmful to health or are generated in the course of any process carried out in the factory.

Where Section 4 of the Factories Act, 1961 might exclude certain areas such as boiler-houses or the provision of breathing apparatus, personnel managers should be reminded that the Health and Safety at Work Act, 1974 will require an employer to take all reasonable steps to ensure the health and safety of the workforce.

Control of dust and fumes

This applies where:

- any dust or fumes or other impurities which are likely to impair health or are regarded as offensive to the employee;
- substantial quantities of dust of any kind are given off.

Employers must take measures to:

- protect employees against the inhalation of dust, fumes and/or other impurities;
- prevent dust from accumulating in the workplace;
- provide a free flow of clean air.

A list of specific regulations referring to dust and fumes is given in Fig. 5.6.

Some specific regulations relating to dust and fumes	Relevance	
	YES	NO
Control of Asbestos at Work Regulations, 1987 Grinding of Metals (Miscellaneous) Regulations, 1925 and 1950 Grinding of Cutlery and Edge Tools Regulations, 1925 and 1950 Blasting (Castings & Other Articles) Special Regulations, 1949 Foundries (Parting Materials) Special Regulation, 1950 Non-Ferrous Metals (Melting and Founding) Regulations, 1962 Chemical Works Regulations, 1922 Indiarubber Regulations, 1922 Chromium Plating Regulations, 1931 Iron & Steel Foundries Regulations, 1953 Highly Flammable Liquids and Liquified Petroleum Gases Regulations, 1972 Control of Lead at Work Regulations, 1980 Factories (Flax and Tow Spinning and Weaving) Regulations, 1906 Factories (Hemp and Jute Spinning and Weaving) Regulations, 1907 Jute (Safety, Health and Welfare) Regulations, 1948 Construction (General Provisions) Regulations, 1961 Smoke Control Areas Regulations, 1990 Continue this list and keep it up to date. It will be checked at safety audit.		

Satisfactory/Unsatisfactory?	Yes/No

Figure 5.6 Some regulations covering dust and fumes

Recognised methods of controlling airborne pollution are given in Fig. 5.7.

NOISE AND VIBRATION

There is a requirement under the Health and Safety at Work Act, 1974 to provide a healthy and safe working environment. It has long been accepted that exposure to severe noise and vibration can seriously threaten a person's health. In addition, the Noise at Work Regulations, 1989 requires:

- employers to prevent damage to hearing;
- designers, manufacturers, importers and suppliers to prevent damage to hearing.

Composition of pure air		At safety audit
Oxygen Carbon Dioxide Nitrogen and other inert gases	20.94% 0.03% 79.03%	

Satisfactory/Unsatisfactory?	Yes/No

Type	Hazardous substance	Source	Process	Possible effect	Possible remedial action
DUST	Silica (Clay) Hardwood Spore	Dry sweeping Sandling Mouldy hay	Pottery Woodwork Agriculture	Silicosis Nasal Cancer Farmer's lung	Dampen down Respirator Ventilation
FUMES	Zinc Cadmium	Hot flame Heat	Flamecutting Hard solder	Fume fever Emphysemia	Ventilation Respirator Extractor
GAS	Nitrogen Oxides Carbon Dioxide Slurry Gases	Hot flames Engine exhaust Fermentation	Welding Garages Agriculture	Lung irritation De-oxygenation Asphyxiation	Ventilation Respirator Extractor
VAPOUR	Perchloroethylene Isocyanate	Evaporation Moulding	Dry cleaning Plastics	Liver damage Asthma	Ventilation Respirator Extractor
MIST	Chromic acid Non-solvent oil Animal infection	Bubbles breaking Machine lubricant Meat handling	Plating Engineering Abattoir	Ulceration Skin cancer Brucellosis	Ventilation Protective Clothing Respirator

At safety audit these are to be examined and form part of the evaluation process

Figure 5.7 Some ways of controlling airborne pollution

Employers are deemed to include self-employed persons and an employee must co-operate with their employer's programme to prevent hearing damage. Within these regulations, every employer shall ensure that a competent person makes a noise assessment which is capable of:

- identifying which employees are exposed;
- providing information with regard to noise exposure levels;

Noise level	Checklist	YES	NO
> 85 dB (A)	Are noisy machines or processes identified by warning signs? Does everyone in the noisy area need to work there? How long can people stay in the noisy area? Have employees been warned about the dangers of noise? Have they ear protectors? Do they wear ear protectors? Has manufacturer's information about noise levels been checked? Will changes in work methods affect noise levels? Can noise be reduced by fixing loose/vibrating pieces? Can better maintenance reduce noise levels?		

Comments:

Action taken:

Signed: Date:

At safety audit these records will be examined

Figure 5.8 A simple noise level checklist

- providing a review procedure of noise levels where necessary.

Noise assessment records should be kept so that trends may be identified over longer periods of time. A sample noise assessment form is given in Fig. 5.8.

Every employer shall undertake a programme of measures to:

- identify noise sources;
- identify remedial measures to be taken;
- implement those remedial measures;
- ensure that action is taken;
- monitor the situation;
- reassess noise exposure levels.

Location	Background Noise Level	Peak Noise Level	Date of Assessment	Action Taken	Person Responsible

These assessment forms will be examined at safety audit

Figure 5.9 A simple noise assessment form

Factors to consider are:

- the number of workers who would benefit by the noise reduction programme;
- the noise exposure levels involved;
- a socio-technical appraisal of noise reduction strategies;
- factors which might impede wearing of ear defenders;

In areas where noise levels require remedial action appropriate signing must be displayed warning that ear protectors must be worn.

There is a need to take action if workers receive a daily personal exposure to noise at or above 85dB(A) or at a peak sound pressure at or above 140dB(A). A checklist is provided in Fig. 5.9.

PART 2

DECISION-MAKING ENVIRONMENT

It is essential that the safety auditor examines the way in which decisions are made within the organisation and the speed with which they are made and implemented. This aspect of the audit would also examine the communication aspects of the safety processes and comment upon ways in which messages, policy changes and so on are made within the company.

The safety auditor should examine the ways in which the safety mix is incorporated into the decision-making process and the methodology employed whilst at the same time examining the safety management processes as defined in the Handbook of Safety Management, p.120.

SAFETY COMMITTEES

A unionised employer has a duty to consult with safety representatives appointed by the appropriate trade union. Where two or more safety representatives require the formation of a safety committee, an employer must undertake the establishment of such a committee under section 2(7) of the HASAWA and the Safety Representatives and Safety Committees Regulations 1977, reg. 9(1) within three months of the request being made. It is important to check company policy on these matters and with trade union liaison in general (see below).

A safety representative is an employee nominated by his trade union to represent his colleagues in discussions with the employer on matters relating to health and safety at work. He may carry out surveys or inspection of the workplace with the object of identifying hazards or potential dangers. Employers are required to disclose to such representatives all information necessary for them to carry out their duties and functions. Details of the information can be found in para. 6 of the Approved Code of Practice on Safety Representatives and Safety Committees. At the same time, health and safety inspectors are required to provide safety representatives with technical information obtained as a result of a visit to their workplace. This must include details of prosecutions, improvement notices, prohibition orders, and any correspondence to the employer on health and safety matters. All action considered by an inspector as the result of a visit must be discussed with the safety representative.

As far as the personnel manager is concerned, the safety representative/s within an organisation are an extremely valuable asset. They must be met regularly, involved directly within the decision-making process and considered an extension of the overall health and safety programme.

For safety committees to be effective they must not be too large otherwise their ability to make decisions effectively is diminished. They must be made up of senior management, trade union safety representatives with the secretariat being provided for by the safety manager or one of his members of staff. Chairmanship of the committee should follow normal electoral procedure and a constitution should allow for periodical changes to the structure. Decisions taken by the committee must be adhered to and carried out as soon as is reasonably practicable. Merely having a safety committee because the regulations say you have been requested to have one is a severe waste of resources if it is not permitted to make an effective contribution to the well-being of the organisation.

Decisions concerning new laws and regulations affecting working practices should be placed before the safety committee for discussion. From this, all matters concerning the planning, implementation, monitoring and evaluation of policy can be discussed openly.

These regulations came into force on October 1, 1978 and gave trade unions the right to appoint safety representatives to perform the following functions:

- To investigate potential hazards and dangerous occurrences at the workplace and to examine the cause of accidents.
- To investigate complaints by employees in relation to health, safety and welfare at work and to make representations to the employer on these issues.
- To take up general matters of health and safety at work and to represent employees in consultation with HSE inspectors.
- To attend meetings of the safety committee when and if necessary.
- To carry out a workplace inspection every three months after giving reasonable notice to the employer of intention to do so.
- To carry out workplace inspections (after consultation with his employers) when there has been a change in conditions of work, or new information has been issued concerning relevant hazards. The employer is entitled to be represented at these inspections if he wishes.

The idea of workers being involved in these activities has been debated for many years and was actually resisted by many trades unions because they suspected that such involvement might relieve employers of some of the responsibility to health and safety.

It is important to note therefore that the activities outlined above are merely functions which the representative may carry out. They do not relieve the employer of any responsibility whatsoever and do not confer any legal liability on the appointed representative.

The regulations specify a number of other provisions relating to the training of safety representatives and the provision of information relating to safety and health at work. Your organisation's safety policy may specify the local arrangements made to enable safety representatives to perform these functions. Safety representatives are an important influence in the provision and maintenance of a safe and healthy working environment.

MANAGEMENT INVOLVEMENT

A note must be made of the management involvement in health and safety issues in the workplace and a record should be made of:

- senior management interest and enthusiasm for health and safety;
- management support within the decision-making process;
- management and supervisory involvement in accident reduction and prevention;
- health and safety – that it is given a high profile in the organisation.

It is essential that management involvement in the safety audit process is closely monitored and forms an integral part of the investigative process.

TRADE UNION LIAISON

Trade unions and industrial relations is a multi-disciplinary subject which concerns both the individual and groups and which should be understood by safety practitioners. It can be regarded as a systematic approach to the analysis of social systems in which workers, trade unions, employers, the state and its agencies interact with one another. These are in turn influenced by economic, technological, social, political, ideological and historical factors in the environment. From the systematic analysis of these factors, an attempt is made to discover the rules and rule-making processes by which they regulate their behaviour and to define what they are all trying to do.

Within this broad framework, it does not necessarily follow that all the participants share a common view of the purpose of their work or the solutions to the industrial relations problems. Some tackle the problem from a different angle. For example, some see it as the development of relations between management, workers, trade unions and employers. Although such groups are usually hard-headed, pragmatic and unsentimental, they involve the use of language complete with normative and moralistic content. The most common application of normative vocabulary in industrial relations involves the concept of fairness: fair wages; fair comparisons; unfair dismissals; unfair industrial practices and a fair day's pay for a fair day's work are common terms amongst industrial relations.

The formation of attitudes towards fairness has been conducted within two perspectives:

1. This approach makes use of the reference group theory which is interpreted within a social psychological frame of reference. This perspective has raised important questions of not only how workers compare their own employment with those of individuals and groups but also why particular orbits or comparison are selected.

2. This method makes use of the industrial relations' institutionalists who refer to the notion of an industrial relations system. Here it is likely that questions regarding social structure, (in the context of collective bargaining) are overlooked, but also this process can fail to recognise the complex interrelationship between the structure of inequality and occupational worth. Any explanation which excludes consideration of these factors amongst the determinants of pay and attitudes to pay can be regarded as incomplete.

In bringing these two approaches together, industrial relations may be seen as a constant process of change where new experiences and the results of empirical research will steer the safety manager away from the opposing assumption that industrial relations and safety management is a static subject. Having said that, the issues of industrial relations arise out of:

– Who does what?
– Who gets what?
– Is it fair?

These questions usually relate to income, treatment and time, and these issues provide potential for the emergence of conflict and hence the problem of resolution and cause about which management needs to be aware.

Income. Aspects regarded as unfair:
– between the low paid and the rest;
– between earned and unearned income;
– between wage earnings and staff salaries and fringe benefits;
– between occupations;
– between grades and internal differentials;
– between public and private sectors;
– between jobs in different regions;
– does the incomes and wages system reflect the changing aspirations of the participants?
– does it accommodate technical change?
– does it enable changes in output?

Time.
- Is overtime necessary?
- Shift work and the conflict between man and machine.
- Is flexi-time a possibility?

Treatment. Are the needs and expectations being catered for concerning the following:
- fringe benefits, staff status, etc.;
- management needs organisation – do workers respond as individuals or as informal working groups?
- management must initiate change whilst the worker may be interested in security;
- the worker and manager are both employees but may also be members of external organisations such as professional associations or trade unions.

The study of industrial relations

It has been said above that the study of industrial relations can be regarded as the study of rules and rule-making processes which regulate the employment relationship. The purpose of rules, both formal and informal exist to establish rights and obligations which together outline spheres of authority and define status and thus establish norms of expected and appropriate behaviour. The main rules and rule making processes are:

- Legislation.
- Collective bargaining.
- Unilateral management decisions.
- Unilateral trade union regulations.
- The individual contract of employment.
- Custom and practice.
- Arbitration awards both voluntary and statutory.
- Social conventions.

Thus, apart from the problems of fairness mentioned earlier as a means of providing potential conflict, aspects of the rule-making processes are also areas of possible controversy:

- Who should make the rules and how?
- What is the right process?
- What issues should the rules cover?
- How should the rules be administered?
- How should the rules be enforced?
- How should the rules be changed?
- How should the rules be legitimised by those affected?
- What effects will the new rules have on existing managerial prerogatives?

The safety manager is concerned with two types of rules which are referred to as substantive and procedural rules.

Substantive rules

1. Those rules governing compensation in all its forms.

2. Rules regulating duties and performance expected from employees, including rules of discipline for breaches of rules or standards.

3. Those rules defining the rights and duties of employees which will include new or laid-off workers, to particular positions or jobs.

Procedural rules

These are concerned with defining procedures for the establishment and administration of the substantive rules given above. In essence, substantive rules define jobs, whilst procedural ones regulate the defining process.

In order to get agreement on fairness it may be necessary to use job evaluation and analysis exercises concerning:

- skill and training;
- social worth;
- risk and safety;
- physical effort;
- experience;
- responsibility.

But other factors are also relevant such as:

- ability and age;
- expected working life of the job;
- productivity;
- persistence, reliability and honesty;
- scarcity value;
- social needs and traditions.

It is important to establish what effect each factor will have. Although job evaluation over an industry may overcome unfairness among people in the workplace and cater for unfairness arising from regional differences, can this method be extended to cover the other sources of unfairness which have been discussed so far? In terms of collective bargaining, fairness is only felt by the parties to the bargain and it tends to preserve an unfair *status quo*. In many ways, it can be said that the role of a trade union here is to act as a pressure group in defence of such differentials. It is not the intention within this chapter to discuss collective bargaining in depth and those wishing to learn more of this procedure should read other books on this subject, some of which are given at the end of the book.

From a safety management point of view, a company's industrial relations policy should form an integral part of the total strategy with which it pursues its business objectives. There are five advantages to this in that it provides an ideal atmosphere for:

- consistency;
- orderly and equitable conduct;
- planning;
- anticipation of events;
- retaining the initiative in changing situations.

Being aware of relevant industrial relations objectives and the establishing of principles and guidelines for management, the areas of important activities to be covered by a framework of objectives and principles are:

- Company responsibilities, management prerogatives and union rights.
- Union recognition and facilities.
- Attitudes towards collective bargaining.
- Bargaining structures.
- Dispute settlement procedures.
- Payment systems.
- Security of employment.
- Communications.
- Employee involvement.
- Disclosure of information.
- Health and safety.

Outline for a typical company industrial relations policy

The objectives of a company's industrial relations policy will differ from organisation to organisation and from the type of industry it operates within. Some companies will express their objectives in quantitative terms as far as possible relating industrial relations objectives to corporate objectives. They would not wish to omit references to behavioural implications. Thus, senior management may consider such possible objectives as the:

- Development of an atmosphere of mutual trust and co-operation at the workplace.
- Prevention of problems and disputes wherever possible.
- Provision of solutions to problems and disputes which arise through agreed procedures.
- Encouragement of opportunities for employee motivation, development of skills and productivity among all categories of employee.

- Reduction or stabilisation of labour costs.
- Strengthening of managerial control over the work situation.
- Reduction or prevention of accidents and health promotion.

A framework for the viable management of industrial relations should contain four essential features:

1. Management accountability. Full acceptance by management for industrial relations in exactly the same way as for product quality or for marketing strategy.
2. Management initiative. Taking the initiative in collective bargaining and other facets of industrial relations.
3. Management distinction. This is based on the concept of accountability between the legitimate function of trade union and management.
4. Management practice. Where policy is a statement of objectives and commitment to principles issued by senior management and its application in practice.

To be effective, an organisation must have a industrial relations policy which conforms to certain standards such as:

- the policy statement must be universally applicable;
- it should be in writing;
- it should be flexible to meet varying local conditions;
- it should be justifiable on an assessment of profit forecasts or other criteria;
- it should be approved and authorised by the company chairman, chief executive or president;
- it must be regarded as inviolate as far as is possible.

It is necessary for employees, line managers, senior managers and personnel/industrial relations specialists to have a basic understanding and knowledge of organisational policy.

A number of organisations are against the involvement of trade unions in company affairs. In terms of safety the more help an organisation can get in reducing or preventing accidents from happening the better. Trade unions, if used correctly by management, can be of great assistance and must form part of the safety audit.

LIAISON WITH OTHER INTERESTED GROUPS

It is important that regular contact is kept with other groups interested in the safety and health of the workforce. This could involve regular contact with such organisations as the Institution of Occupational Safety and Health, the

International Institute of Risk and Safety Management, the Health and Safety Executive, the Royal Society for the Prevention of Accidents, the Institution of Environmental Health Officers, the Royal Society of Health, the British Safety Council and so on. These organisations are in the forefront of the profession and are a valuable source of information, guidance and advice. The safety auditor should examine the influence these organisations have on the company being audited and a note made of the publications and other advice which has been obtained and used.

DISASTERS AND EMERGENCY PLANS

To prevent accidents from occurring, there are a series of laws and voluntary strategies which can be implemented cheaply and effectively. This book and those suggested for *further reading* suggest ways in which various accident prevention and reduction strategies can be planned, implemented, monitored and evaluated. A disaster is no different and can thus be prevented using the same principles and practice. However, there are varying degrees of disaster from the straight forward typical workplace accident to one which can cause the workplace to cease trading completely or cause great harm to people and property outside the immediate vicinity of the incident. It is the latter type of incident which must be considered here. Very few organisations possess emergency plans which can be brought into operation and this small contribution merely serves to prompt personnel managers to seek appropriate advice by obtaining answers to the following:

- Do you have criteria set out which describe clearly a disaster to your company?
- Do you have a procedure to contain the disaster?
- Do you have plans to enable you to continue trading?
- Do you have a building evacuation procedure?
- Are appropriate personnel trained in your emergency procedures?
- Have you tested your emergency plans?
- Are your plans adequate?
- Have you publicised your prevention policies, procedures and practice to the workforce?

These are a sample of the basic questions which must be asked. The most common type of disaster is a fire and this matter is covered in more detail in Chapter 6. It must be remembered that although adequate insurance may be held by an organisation, it must have some plans to continue trading safely in the event of a disaster. Insurance alone is not enough.

STATEMENTS OF HEALTH AND SAFETY POLICY

One of the legal provisions of section 2 of the HASAWA requires employers who employ five people or more to prepare, and keep up to date, a written statement of their policy regarding the health and safety of their employees.

Your policy may consist of one complete manual or it may be a compilation of several individual documents relating to particular areas of your activity. Whatever form it takes it should comprise three separate and distinct parts:

1. General statement of policy.
2. Organisation and responsibilities for carrying out the policy.
3. Arrangements for ensuring safety and health of employees.

Many organisations produce a general statement of intent and require individual departments, cost centres or directorates to develop safety policies which appropriately reflect the day-to-day activities of the department. By looking through the policy you should find an indication that the organisation will provide resources for health and safety along with the provision of safe plant and equipment, safe systems of work, training for staff and supervision. In fact, you should observe a resemblance to the general duties laid down by section 2 of the HASAWA.

As far as responsibilities are concerned, these should include the duties of employees at all levels in the organisation.

Section 1 General Statement of Policy
Here you should state policy in respect of:

*   provision and maintenance of safe and healthy working conditions;
*   equipment and safe systems of work for all employees;
*   provision of health and safety information, training and supervision necessary;
*   a statement regarding the acceptance of responsibility for the health and safety of other people who may be affected by your activities.

Within this section, it is good practice to outline the allocation of duties for all safety matters and the particular arrangements which will be made to implement your company health and safety policy. Your statement of health and safety policy should always reflect the changes in nature and size of the company and as such be kept up to date. To ensure this, personnel managers responsible for health and safety matters within their organisations should see to it that such statements are reviewed every year and amended accordingly.

Section 2 Organisation and responsibilities for carrying out the policy
This section should cover those general arrangements for carrying out health and safety policy such as:

- procedures for the reporting and recording of accidents and dangerous occurrences;
- location of first-aid boxes;
- details of all qualified first-aid personnel;
- details of fire procedures;
- fire extinguisher location and maintenance details;
- fire alarm location and maintenance details;
- fire routines and testing arrangements;
- medical and health care arrangements;
- safety training details including specialist training;
- discipline and codes of conduct;
- advice for visitors and/or contractors.

Section 3 Arrangements for ensuring safety and health of employees

This final section should list details concerning all hazards in the workplace. These should be listed so that everyone is aware of them. If you use hazardous substances, then the manufacturer will provide hazard sheets, whilst the Health and Safety Executive (HSE) will provide additional advice if it is required. Make sure that the workforce is aware of all hazards. This section should also provide information concerning:

- cleanliness of the premises;
- waste disposal details;
- details concerning safe stacking and storage;
- marking and maintaining clear walkways and exits;
- equipment checking procedures;
- details concerning access to restricted areas;
- routines for checking electrical appliances;
- routines for the reporting of faults;
- rules regarding the use of extension cables and portable equipment;
- arrangements with electrical and/or equipment contractors;
- rules concerning the use of all machinery;
- details of routine maintenance and timetables;
- dangerous substance details;
- protective equipment and clothing information;
- routines for storing, handling and disposal of dangerous substances;
- operation, use and maintenance of compressed air equipment;
- storage, labelling and use of compressed gases and/or fluids;
- procedures regarding water pressure and/or steam;
- in-house rules and regulations such as:
  - use of internal transport;
  - use and care of protective equipment and clothing;
  - noise;

- maintenance of appliances;
- any other special hazards in your company.

It is good practice to include a company's health and safety policy in the induction course programme for new employees. A sample health and safety policy document is given in Fig. 5.10.

Health and Safety at Work Act, 1974

This is the policy of:

ABC Manufacturing Ltd
Weymouth Avenue
New Town
Barchester BB1 1AA

Dear Employee

Health and Safety Policy

You will know from your Contract of Employment that you are an employee of ABC Manufacturing Ltd whose health and safety policy applies to you under the Health and Safety at Work Act, 1974.

This document contains a copy of that safety policy and the statement of organisation and other arrangements adopted by the company to implement it.

These documents set out the procedures which management will apply in the interests of the health and safety of all employees. It is our wish to emphasise that each and every one of us has a duty to take reasonable care of ourselves and our fellow workers.

We must all work together to prevent accidents happening and the hardships that they cause.

Yours sincerely

Chairman (or Managing Director)   January 1993

This introduction may be in the form of a letter on company headed paper to each employee together with the policy document or bound in as an integral part of the document.

Figure 5.10 A sample health and safety policy document

ABC Manufacturing Ltd

Health and Safety at Work Act (HASAWA) 1974

The purpose of this document is to set out the health and safety policy of ABC Manufacturing Ltd as required by section 2(3) of the Health and Safety at Work Act 1974.

General Policy

It is the policy of ABC Manufacturing Ltd to:

1. safeguard the health, safety and welfare of all its employees while at work and to provide, as far as is reasonably practicable, working environments which are safe and without risk to health;

2. conduct its undertakings in such a way as to ensure, so far as is reasonably practicable, that people not in its employment, but who may be affected, are not exposed to risks to their health and safety;

3. recognise its obligations to meet all relevant legislative requirements pertaining to health and safety which apply to any of the company's undertakings;

4. organise and arrange its company affairs to ensure compliance with this policy.

In carrying out this policy, it is the practice of ABC Manufacturing to:

1. specify in writing managerial responsibility and accountability for the health, safety amd welfare of its employees and for the health and safety of others who may be affected by the company undertaking;

2. ensure appropriate safety training and instruction for new and/or unfamiliar methods and equipment, and that accident prevention is included in all relevant training programmes, especially for apprentices, young trainees, new members of staff and other employees;

3. use appropriate propaganda and other relevant measures to ensure an awareness of the need to prevent accidents and risk to health in the minds of employees;

4. take into account, when planning its work, any aspects which will help to eliminate injury, industrial disease, pollution and waste;

5. make appropriate accident prevention arrangements at the place of work and maintain liaison with all other employers who have employees working at the same location;

6. encourage the discussion of health and safety matters at all levels within the company, including the setting up of arrangements for joint consultation with employees, through their appointed safety representatives.

The Health and Safety at Work Act 1974

The company recognises that it has a legal requirement to bring to the notice of all its employees its health and safety policy. The company will maintain contact with the Health and Safety Executive (HSE), its own consultants and other groups in order to keep itself informed about health and safety matters. It also recognises that there exists a large body of authoritative documentation in the form of legislation, approved codes of practice, HSE guidance notes, HSE publications and British Standards. ABC Manufacuring Ltd will establish and maintain an index of all such relevant publications so that they may be readily identified by the company and its employees.

Figure 5.10 cont. A sample health and safety policy document

ABC Manufacturing Ltd   Appendix 1

Health and Safety at Work Act, 1974

RESPONSIBILITIES:

Overall responsibility for Health and Safety in ABC Manufacturing Ltd is:

Mr Albert Bloggs, Managing Director

Responsibility for the implementation and monitoring of this policy is the responsibility of:

Mrs Betty Lepew, Works Manager

In the absence of the Works Manager the person below is responsible:

Mr James Tuner

The following supervisors are responsible for health and safety in specific areas:

Mrs Alice Springs	Finishing Room
Mr Mal Coombe	Tool Room
Mrs Irene Box	Despatch
Mr Uriah Heap	Stores
Mrs Felicity Harper-Brown	Sales
Mrs Jane Scudd	Administration

All employees have the responsibility to cooperate with managers and supervisors to achieve a healthy and safe workplace and to take reasonable steps to care for themselves and others.

If you notice a health and safety problem, and cannot rectify the matter then you must inform one of the persons whose names are shown above as soon as possible.

Consultation between management and employees is provided by:

Mr Fred Smallpiece
National Union of Widget Tappers and Nut Nobblers

The following staff are responsible for:

Safety Training	Mrs Wendy Baggs
Safety Inspections	Mr Runjit Patel
Accident Investigation	Mrs Wendy Baggs
Safety Maintenance	Mr James Dyer

Figure 5.10 cont. A sample health and safety policy document

ABC Manufacturing Ltd   **Appendix 2**

Health and Safety at Work Act, 1974

First-Aid Boxes are sited at:

> Finishing Room
> Tool Room
> Despatch
> Stores
> Sales
> Administration

Those responsible for their first-aid boxes are the following qualified first aiders:

Finishing Room	Winston Bakerlight
Tool Room	Crispin Smith
Despatch	Tom Cobley
Stores	Anne Seagrove
Sales	Alison Baker-Brown
Administration	Mary Grimshaw

Mrs Jane Scudd of Administration is the Company Fire Officer. Her telephone number is:

> Ext 4550

Mrs Scudd is responsible for evacuation procedures and for the maintenance of all fire equipment. She, from time to time (at least once per month), tests the fire alarms. Escape routes and assembly points are clearly shown on the diagram displayed on the notice board in your work area.

Hosepipes and extinguishers are maintained by:

> Firehose and Fire Extinguishers Ltd
> New Town
> Barchester BB3 3AC
> Tel: 0333 123

Smoke detectors are maintained by:

> Smoke Detectors Ltd
> New Town
> Barchester BB2 6AJ
> Tel: 0333 212

KEEP ESCAPE ROUTES CLEAR

Figure 5.10 cont. A sample health and safety policy document

ABC Manufacturing Ltd

Health and Safety at Work Act, 1974

This company employs the following person as a Safety Consultant:

> R A Blogg, MSc, PhD, MIIRSM, MIOSH, MRSH
> Bloggs Safety and Health Ltd
> New Town
> Barchester BH2 2EZ

The Company Doctor is:

> Dr J Smith, MB, BCh, MRCP
> The Surgery
> New Town
> Barchester BB1 1TY

The Company Nurse is:

> Ms Gladys Emmanuelle, RGN
> The Surgery
> New Town
> Barchester BB7 7TY

Health and Safety Training is provided by:

> Barchester Safety Training Services Ltd
> The Industrial Estate
> New Town
> Barchester BB3 3LL

The following jobs are hazardous and must not be undertaken without specialist training:

> Widget washing
> Nut nobbling
> Tool tapping
> Blank blasting
> Sheet shearing

These jobs use chemicals and equipment which if wrongly used can cause severe accidents and ill health. See your supervisor for specialist training arrangements.

Figure 5.10 cont. A sample health and safety policy document

ABC Manufacturing Ltd   **Appendix 4**

Health and Safety at Work Act, 1974

Outline your procedures to be followed by Contractors and visitors below:

Rules for contractors:

Rules for visitors:

Figure 5.10 cont. A sample health and safety policy document

ABC Manufacturing Ltd   **Appendix 5**

Health and Safety at Work Act, 1974

List below the number of types of hazard which you may need to provide for. Set out the rules.

Hazard sheets:

Manufacturer's guidance:

Other guidance:

House rules:

Figure 5.10 cont. A sample health and safety policy document

ABC Manufacturing Ltd **Appendix 6**

Health and Safety at Work Act, 1974

For housekeeping you will need to list the areas of concern and include the rules involved:

Cleanliness (Buildings):

Cleanliness (Personal):

Waste disposal:

Safe stacking and storage:

Marking and keeping clear gangways and exits:

Checking of equipment and access to special places:

Figure 5.10 cont. A sample health and safety policy document

ABC Manufacturing Ltd   **Appendix 7**

Health and Safety at Work Act, 1974

Electrical equipment will require regular maintenance and checking.
State below:

> **Routines for the inspection of plugs, cables, loose connections and faults:**

> **Rules for using extension cables and portable equipment:**

> **Periodic check arrangements with contractors and/or suppliers:**

Figure 5.10 cont. A sample health and safety policy document

ABC Manufacturing Ltd   **Appendix 8**

Health and Safety at Work Act, 1974

Rules covering the use of all machinery must be set out below:

Rules for the use of:

Widget washer:

Nut nobbler:

Tool tapper:

Blank blaster:

Sheet shearer:

Equipment type:	Who should check it	When should it be checked
Widget washer	Clinton Samual	Daily before each shift
Nut nobbler	Angus McGreggor	Daily before each shift
Tool tapper	Lionel Wilson	Weekly
Blank blaster	Abdul Rashid	Weekly
Sheet shearer	Alan Jones	Monthly

Figure 5.10 cont. A sample health and safety policy document

ABC Manufacturing Ltd   Appendix 9

Health and Safety at Work Act, 1974

Rules concerning the use of dangerous substances should be listed below:

Rules concerning pressurised fluids and their use should be listed below:

Figure 5.10 cont. A sample health and safety policy document

ABC Manufacturing Ltd   **Appendix 10**

Health and Safety at Work Act, 1974

You may need additional rules to cover such matters as:

- internal transport;
- use and care of protective equipment and clothing;
- ventilation;
- noise;
- use of VDUs;
- temperature;
- lighting;
- maintenance of appliances and other hazards connected with your particular business.

Hazards:

Rules:

Maintenance:

Figure 5.10 cont. A sample health and safety policy document

ABC Manufacturing Ltd   Appendix 11

Health and Safety at Work Act, 1974

You may need additional rules to cover such matters as:

Alcohol advice
Drugs and medicine
Welfare issues (such as sports facilities and personal counselling)

Rules concerning alcohol and general advice:

Advice concerning the use of drugs and medications whilst carrying out certain tasks:

Welfare issues:

Counselling and advice:

Other welfare issues such as sporting and leisure facilities:

Figure 5.10 cont. A sample health and safety policy document

ABC Manufacturing Ltd   Appendix 12

Health and Safety at Work Act, 1974

Finally, you may also need rules and advice in relation to road safety such as:

Car parking arrangements
Speed limits on works roadways
Advice on entering and leaving the works
Restricted parking areas
Pedestrian walkways
Works vehicles
General road safety advice

Rules concerning vehicle use:

Parking arrangements:

Security:

Advice concerning own vehicle:

Advice when using works vehicles:

Pedestrian walkways:

General road safety advice:

All these items will be checked at safety audit for relevance and currency.

Figure 5.10 cont. A sample health and safety policy document

ACCIDENTS AND DANGEROUS OCCURRENCES

The requirement to report injuries, diseases and dangerous occurrences are contained in the *Reporting of Injuries, Diseases and Dangerous Occurrences Regulations, 1985*. These are sometimes referred to as RIDDOR for short. These regulations require that should any of the following events occur, the responsible person must report it in writing to the enforcing authority and must keep a record of it. The regulations also stress that the responsible person should also notify the enforcing authority by telephone as soon as is reasonably practicable where:

- a death occurs to any person, whether or not he or she is at work, as the result of an accident arising out of, or in connection with, work;

- any person suffering a specified major injury or condition as a result of an accident arising out of, or in connection with, work;

- one of a list of specified dangerous occurrences arising out of, or in connection with, work;

- a person at work being incapacitated for his or her normal work for more than three days as a result of an injury caused by an accident at work;

- the death of an employee if this occurs some time after a reportable injury which led to that employee's death, but not more than one year afterwards;

- a person at work being affected by one of a number of specified diseases, provided that a medical practitioner diagnoses the disease and that the person's job involves a specified work activity.

A diagram illustrating the procedure to be followed is given in Fig. 5.11.

When making a report, it is necessary to complete Form 2508 (revised January 1986) for the reporting of deaths, injuries and dangerous occurrences and Form 2508A should be used when reporting cases of disease.

The regulations also require employers to keep records of every event which has caused injury, however small, or which has the potential to cause injury. The details to be kept are as follows:

- Date and time of the accident or dangerous occurrence.
- Details of the person(s) affected including:
  - full name;
  - occupation;
  - nature of the injury or condition.
- Location where the accident or occurrence took place.
- A brief description of the incident.

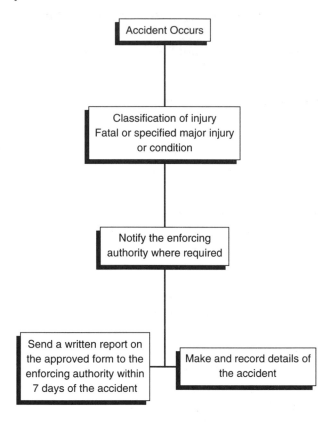

Figure 5.11   Report and record accidents which happen

In addition, where a disease is concerned the following information should be recorded:

- date of diagnosis;
- personal details of affected person;
- name and nature of the disease.

All records must be kept for three years and be made available to the enforcing authority if or when required. There is flexibility in the way records will be kept provided that they contain the above particulars. An example form is given in Fig. 5.12.

ABC Manufacturing Ltd   Ref Number

Name

Age

Address

Works Number

Employed as

Post code

Date commenced

STATEMENT – concerning the accident

Date of accident   Time of accident   Number of
                                                     people involved

Location of accident

Details of injuries received   Was this a fatal accident YES/NO

Is this the statement of: THE CASUALTY/WITNESS?

Statement:

Continue on a separate sheet if necessary

All accident records will be examined at safety audit.

Figure 5.12 Witness or casualty record sheet

LIABILITY INSURANCE

It is a statutory requirement for most companies in the United Kingdom to have an insurance policy which provides cover against claims for injury and/or disease by employees. When a policy is taken out, the insurers provide a certificate of insurance which the employer must display in a prominent position at the work site. This is so that employees and anyone else requiring to see it can do so. It is a criminal offence not to take out such insurance and/or to display the certificate.

The object of having such insurance is to make certain that employers are covered for any legal liability to pay damages to employees who are injured or caused ill health through their employment. Such a policy protects an employer from third party claims but not for non-employees. Where contractors or members of the public are concerned it is necessary to take out additional insurance referred to as public liability insurance. Whilst such a policy might be desirable it is not a requirement of the law. All statutory liability insurance must be available under one or more approved policies. These policies are not subject to any conditions or exceptions prohibited by the law.

An employer is not required to insure against liability to an employee who is:

- a father;
- a mother;
- a son or daughter;
- a wife or husband;
- a close relative.

With regards to the degree of cover necessary, it is a requirement to maintain insurance of £2,000,000 but most reputable insurance companies have no upper limit.

Exempted employers are:

- local authorities;
- civil service;
- nationalised bodies;
- commission for the new towns;
- statutory water authorities;
- London Transport.

Failure to effect and maintain valid insurance for any day on which it is required is a criminal offence as is failure to display a valid certificate of insurance.

WOMEN WORKERS

There are many books available covering aspects of the Sex Discrimination Act and it is not intended to recover this ground here. In terms of health and safety however, case law has shown that a woman's health and safety in certain circumstances takes precedence over a duty not to discriminate. For example, in the case of *Page v. Freight Hire (Tank Haulage) Ltd [1981] IRLR 13* a 23-year-old woman was employed as a heavy goods vehicle driver of tankers carrying the chemical dimethylformamide (DMF). On a recommendation from ICI that women of child-bearing age should not be employed in driving lorries loaded with the chemical DMF, Mrs Page was taken off this kind of work. She brought a claim of unlawful sex discrimination. The decision did not allow an employer to stop employing women as soon as a hint of danger arose. What needed to be considered was the employer's duty under section 2 of the Health and Safety at Work Act. In this particular case, evidence from the manufacturer was that the chemical was dangerous to women of Mrs Page's age and the steps taken to remove her from the danger was a reasonable one. There was, therefore, no unlawful discrimination.

The effect of the Sex Discrimination Acts of 1975 and 1986 has been to remove some of the restrictions on women in relation to their work but restrictions and prohibitions on certain types of employment are still in force in the interests of health and safety.

INDUSTRIAL TRIBUNALS

Industrial relations is a multi-disciplinary subject, the content of which is still in controversy. This is not surprising because the employment relationship is complex in so far as it is essentially an economic relationship which takes place within a:

- social
- psychological
- political
- cultural

context, expressed legally in an individual contract of employment. The background to this relationship is often a complex pattern of shifting forces and changing environments, public pressures, private fears, world markets, weekly household bills, driving ambitions and security. The issues involved with industrial relations arise out of who does what? Who gets what? What is fair in relation to:

- income;
- treatment;
- time?

These issues provide potential for the emergence of conflict and thus the problems of resolving conflict and its causes are issues of which management must be aware.

Income
Aspects regarded as unfair:

- between the low paid and the rest;
- between *earned* and *unearned* income;
- between wage earnings, staff salaries and fringe benefits;
- between occupations and between industries;
- between grades and internal differentials;
- between public and private sector;
- between jobs in different regions;
- does the wage system reflect the changing aspirations of the staff?
- does it accommodate technical change?
- does it enable changes in marketing and sales policies?

Treatment
Are the needs and expectations of the modern industrial worker being catered for in respect of:

- fringe benefits;
- staff status;
- facilities.

Management needs organisation but workers respond as individuals or as informal working groups. Whilst management might initiate change, the worker may only be interested in security.

Time
- is overtime necessary?
- shift work, the conflict between man and machine?
- stress?
- bonus schemes?

The time factor has been identified as having a strong association with accident causation.

There are a number of *rule-making* processes regulating the employment relationship. The purpose of rules, both formal and informal, exist to establish 'rights' and 'obligations' which together outline spheres of 'authority' and define 'status' thus establishing 'norms' of expected and appropriate behaviour. The main rules and rule-making processes are:

- legislation;
- collective bargaining;
- unilateral management decisions;
- unilateral trade union regulations;
- the individual contract of employment;
- custom and practice;
- arbitration awards both voluntary and statutory;
- social conventions.

Thus, apart from the problem of *fairness* providing potential for conflict, so also do aspects of the rule-making processes. For example:

- Who should make the rules and how?
- What is the appropriate process?
- What issues should the rules cover?
- How are the rules to be implemented?
- How are the rules to be enforced?
- How are the rules changed?
- How are the rules to be legitimised by those affected?
- What are the effects of new rules on existing managerial prerogatives?

The two types of rule are substantive and procedural as discussed earlier in this chapter.

Substantive rules define jobs, whilst procedural rules regulate the defining process. Getting agreement on what is regarded as being fair would be the use of job evaluation in order to:

- assess skill and training requirements;
- assess the social worth of the job;
- assess the physical effort required;
- assess the level of experience and competence required;
- assess the level of responsibility;
- assess the level of danger and risk.

Other relevant factors are:

- ability and age;
- working life expectancy of the job;
- ability of the employer to pay;
- productivity;

- persistence, honesty and reliability;
- scarcity value;
- social needs and traditions.

Where conflict occurs there must be a means of resolving the matter quickly and safely. If the matter cannot be resolved internally then industrial tribunals have been set up to consider cases where independent arbitration is necessary. They were originally intended to deal with disputes and injustices in the employment sphere quickly, cheaply and informally. To all intents and purposes, the industrial tribunal is made up of a panel of three persons of which the chairman is usually the only one who is legally qualified. Whilst there is no need to be represented by a legal expert the advice today is that the employment law is so complicated it is often recommended that legal representation or legal aid be provided. Where an employee is a member of a trade union it is usual for the trade union to represent their client.

The make up of a tribunal is usually one trade union panel member, one employer and the chairman who ensures that the complexities of the law are interpreted and followed. For a decision to be reached, a tribunal needs a majority view. Decisions may be challenged provided that:

- the decision discloses an error on the part of the tribunal;
- a party did not receive notice of proceedings;
- a decision was reached in the absence of a party entitled to be heard;
- new evidence has come to light;
- the interests of justice require a review.

The chairman can refuse an application for a review if he believes it will be unsuccessful. Applications for reviews must normally be made within 14 days of the decision of the tribunal being made known.

Industrial tribunals are empowered to hear complaints under the following Acts:

- Equal Pay Act, 1970
- Health and Safety at Work Act, 1974
- Sex Discrimination Act, 1975
- Trade Union and Labour Relations Act, 1976
- Race Relations Act, 1976
- Employment Protection (Consolidation) Act, 1978
- The Employment Act, 1980
- The Employment Act, 1982
- Sex Discrimination Act, 1986

and subsequent similar legislation. Some examples of health and safety matters are:

- failure of an employer to pay safety representatives for time off for carrying out duties and for training;
- failure of an employer to make a medical suspension payment;
- appeals against improvement and prohibition notices;
- time off for training of safety representatives;
- dismissal, whether actual or constructive, following a breach of health and safety law, regulation or term of an employment contract.

Appeals are made against improvement notices and prohibition notices on the following grounds:

- substantive law involved;
- time limits for compliance.

Matters regarding grievance procedures, negotiation, bargaining and the study of industrial relations will be covered in more detail in other books in this series.

Application	Required Protection	Regulations
Workplaces where bottles and syphons are manufactured	Aprons Face guards Footwear Gauntlets	Aerated Water Regulations, 1921
Workplace where asbestos is used	Breathing apparatus Head covering Overalls	Control of Asbestos at Work Regulations 1987
Workplaces where construction and works of an engineering nature are carried out	Eye protectors Helmets Gloves Clothing Belts	Construction Regulations 1961 and 1966
List those carried out by your organisation	List those items of protection you must provide	List those regulations which relate to your industry

These items will be examined at safety audit

Figure 5.13 Example list of protection requirements

SAFETY CLOTHING

Statutory requirements for the provision of protective clothing are laid down by several regulations made under the Factories Act, 1961. Although no specific requirement is made under the HASAWA regarding protective clothing/equipment it does state that employers cannot charge for providing it. Not only can the employer not charge for it but such equipment must be readily available and accessible for immediate use. Where such equipment or clothing is provided an employer must also ensure that it is used or worn correctly.

Within the general terms at common law, the duty of care which every employer owes his workforce requires that he must protect his employees from the risk of reasonably foreseeable injury. Whether a hazard is regarded as reasonably foreseeable must depend upon the circumstances and whether this type of incident has been experienced before. In Fig. 5.13, a summary is provided which contains the requirements for protective clothing in some well known industries and illustrates which regulations refer to each requirement.

It is important that safety equipment and clothing conform to the appropriate Safety Standard and these are published by the Health and Safety Executive. Details are provided in the supplementary reading list at the end of this chapter. Standard personal protection items would include:

- safety eye glasses;
- footwear;
- hats;
- face shields;
- gloves;
- leggings;
- waterproofs;
- ear protection;
- neck protection;
- respiratory equipment.

If you provide such items you must ensure that they are worn correctly at all times.

SUMMARY

It is important that company safety policy is examined thoroughly. It might be necessary to compile an individual list of your company safety policy statements and these should be drawn up in such a way that they can be easily

Safety policy activity	Satisfactory	Not satisfactory
Obtain a copy of the company safety policy document Check currency of company safety policy document Check company insurance documentation Check accident records Check dangerous occurrence reports Check first-aid treatment records Examine legal requirements relevant to the organisation Obtain safety objective statements from: Departments Sections Groups/Teams Individuals Obtain copies of the budgetary provision Examine safety budget expenditure Check priorities for action Examine staff abilities to contribute to safety objectives Examine task analysis data Examine role analysis data Examine safety performance standards data Examine organisational structure for evidence of: Undermanning Poor shift cover Absenteeism Overtime hours Sickness records Obtain safety committee minutes Obtain trade union involvement data Obtain data from safety representatives Obtain safety disciplinary hearing information Examine records concerning: Heating Lighting Noise and vibration Ventilation First-aid provision COSHH data Examine availability of safety literature Examine access to safety literature by employees Examine the condition of health and safety literature and posters		

This is a summary sheet only. Detailed records on each item will also acompany this sheet.

Figure 5.14 An example of a safety audit policy checklist

cross checked in the form of a checklist. Such a list is given in Fig. 5.14 although it must be remembered that such a form should be developed by the safety manager for use in his particular company after taking stock of individual requirements.

From this it is then possible to consider the safety procedures which operate within an organisation.

6 AUDITING SAFETY PROCEDURES

In this chapter we will look at the following:

- **Administrative structure**
- **Communication**
- **Time management**
- **Internal and public relations**
- **Recruitment**
- **Safety training**
- **Supervision**
- **Discipline**

ADMINISTRATIVE STRUCTURE

This aspect of the safety audit should examine the structure of the administrative support which is allocated to the health and safety function. It is important that such administration is seen to serve the health and safety function rather than the other way around. Quite often, on safety audits it can be discovered that some health and safety procedures have to follow the administrative procedures. This should not be the case and all procedures, such as those required under Reporting of Injuries, Diseases and Dangerous Occurrences Regulations, 1985 (RIDDOR) for example, should have the appropriate administrative procedures designed to fit the requirements of the regulations. In this particular case, the RIDDOR requires certain types of information to be sent to external agencies within an appropriate time scale. It is essential, therefore, that the administrative procedures are able to meet this part of the regulations. In addition, other regulations, such as the Dangerous Substances (Notification and marking of sites) Regulations, 1990 also require administrative action and support. It is useful to draw up a list of, say, regulations, (see the example given in Fig. 6.1 below) and check off those which require specific administrative procedures being in place.

Personnel responsible for carrying out the safety audit will require this information in order to test the efficiency and effectiveness of the administrative structure.

Health and safety legislation also places conditions on the level and type of administrative procedures which are in place. For example, in the previous

Some statutory instruments, rules and orders	Does it apply to you?	Have you a copy?	Have you read it?	Do you understand it?
	Yes/No	Yes/No	Yes/No	Yes/No
Factories, Locomotives and Sidings Regulations, 1906				
Factories (Horsehair Processes) Regulations, 1907				
Factories (Electrical Energy) Regulations, 1908				
Tanning (Two-bath Process) Welfare Order, 1918				
Fruit Preserving (Welfare) Order, 1919				
Laundry Workers (Welfare) Order, 1920				
Celluloid (Manufacturing) Regulations, 1921				
Chemical Works Regulations, 1922				
Electric Accumulator Regulations, 1925				
Herring Curing (Scotland) Welfare Order, 1926				
Bakehouses (Welfare) Order, 1927				
Oil Cake Welfare Order, 1929				
Cement Works (Welfare) Order, 1930				
Sugar Factories (Welfare) Order, 1931				
Sanitary Accommodation Regulations, 1938				
Cinematograph Film Stripping Regulations, 1939				
Electricity (Factories Act) Special Regulations, 1944				
Patent Fuel Manufacture (Health and Welfare) Regulations, 1946				
Clay Works (Welfare) Regulations, 1948				
Dry Cleaning Special Regulations, 1949				
Pottery (Health & Welfare) Regulations, 1950				

Figure 6.1 Some statutory regulations

Some statutory instruments, rules and orders	Does it apply to you?	Have you a copy?	Have you read it?	Do you understand it?
	Yes/No	Yes/No	Yes/No	Yes/No
Factories (Testing of Aircraft Engines) Regulations, 1952				
Iron and Steel Foundries Regulations, 1953				
Petroleum Spirit (Conveyance by Road) Regulations, 1957				
Agriculture (Avoidance of Accidents to Children) Regulations, 1958				
Agriculture (Safeguarding of Workplaces) Regulations, 1959				
Agriculture (Threshers & Balers) Regulations, 1960				
Construction (General Provisions) Regulations, 1961				
Agriculture (Field Machinery) Regulations, 1962				
Washing Facilities Regulations, 1964				
Examination of Steam Boilers Regulations, 1964				
Power Presses Regulations, 1965				
Construction (Working Places) Regulations, 1966				
Construction (Health Welfare) Regulations, 1966				
Carcinogenic Substances Regulations, 1967				
Offices, Shops and Railway Premises (Hoists & Lifts) Regulations, 1968				
Asbestos Regulations, 1969				
Abrasive Wheels Regulations, 1970				
Foundries (Protective Footwear & Gaiters) Regulations, 1971				
Highly Flammable Liquids & Petroleum Gases Regulations, 1972				

Figure 6.1 cont. Some statutory regulations

Some statutory instruments, rules and orders	Does it apply to you?	Have you a copy?	Have you read it?	Do you understand it?
	Yes/No	Yes/No	Yes/No	Yes/No
Organic Peroxides (Conveyance By Road) Regulations, 1973				
Agriculture (Tractor Cabs) Regulations, 1974				
Industrial Tribunals (Improvements and Prohibitions) Regulations, 1974				
Woodworking Machines Regulations, 1974				
Protection of Eyes Regulations, 1974				
Employers' Health & Safety Policy Statements Regulations, 1975				
Protection of Eyes (Amendment) Regulations, 1976				
Safety Representatives & Safety Committees Regulations, 1977				
Health & Safety (Enforcing Authority) Regulations, 1977				
Motor Vehicles (Construction & Use) Regulations, 1978				
Control of Lead at Work Regulations, 1980				
Safety Signs Regulations, 1980				
Dangerous Substances (Conveyance by Road) Regulation, 1981				
Diving Operations at Work Regulations, 1981				
Health & Safety (Dangerous Pathogens) Regulations, 1981				
Health & Safety (First-Aid) Regulations, 1981				
Notification of Installations Handling Hazardous Substances Regulations, 1982				
Notification of New Substances Regulations, 1982				
Asbestos (Licensing) Regulations, 1983				
Criminal Penalties (Increase) Order, 1984				

Figure 6.1 cont. Some statutory regulations

Some statutory instruments, rules and orders	Does it apply to you?	Have you a copy?	Have you read it?	Do you understand it?
	Yes/No	Yes/No	Yes/No	Yes/No
Classification, Packaging & Labelling of Dangerous Substances Regulations, 1984				
Poisonous Substances in Agriculture Regulations, 1984				
Social Security (Industrial Injuries) (Prescribed Diseases) Regulations, 1985				
Reporting of Injuries, Diseases & Dangerous Occurrences Regulations, 1985				
Ionising Radiation Regulations, 1985				
Industrial Tribunals (Rules of Procedure) Regulations, 1985				
Building Regulations, 1985				
Electrically Operated Lifts (EEC Requirements) Regulations, 1986				
Dangerous Substances in Harbour Areas Regulations, 1987				
Control of Asbestos at Work Regulations, 1987				
Control of Substances Hazardous to Health Regulations, 1988				
Pneumoconiosis (Workers Compensation) Regulations, 1988				
Classification, Packaging and Labelling (Amendment) Regulations, 1989				
Pressure Systems and Transportable Gas Containers Regulations, 1989				
Public Service Vehicles (Temporary Driving Entitlement) Regulations, 1989				
Road Traffic (Carriage of Explosives) Regulations, 1989				
Road Vehicles Lighting Regulations, 1989				
Freight Containers (Safety Convention) Regulations, 1989				

Figure 6.1 cont. Some statutory regulations

Some statutory instruments, rules and orders	Does it apply to you?	Have you a copy?	Have you read it?	Do you understand it?
	Yes/No	Yes/No	Yes/No	Yes/No
Road Traffic Accidents (Payment for Treatment) Regulations, 1990				
Social Security (Industrial Injuries & Diseases) Misc. Provisions, 1990				
Collision Regulations (Seaplanes) (Amendment) Order, 1990				
Dangerous Substances (Notification and Marking of Sites) Regulations, 1990				
Smoke Control Areas (Authorised Fuels) Regulations, 1990				
Personal Injuries (Civilians) Regulations, 1990				
Cosmetic Products (Safety) (Amendment) Regulations, 1990				
Fire Safety & Safety at Places of Sport Act 1987 (Commencement Order No 6) Order, 1990				
The Agriculture (Tractor cabs) (Amendment) Regulations, 1990				
Dangerous Substances (Notification and Marking of Sites) Regulations, 1990				
Diving Operations at Work (Amendment) Regulations, 1990				
Gas Safety (Installation & Use) (Amendment) Regulations, 1990				
Nuclear Installation Act 1965 (Repeal and Modifications) Regulations, 1990				
Control of Explosives Regulations, 1991				
Packaging of Explosives for Carriage Regulations, 1991 (March 1992)				

Keep this list up to date

Figure 6.1 cont. Some statutory regulations

chapter examples were given in relation to policies concerning heating, lighting, ventilation and so on. These policies will require administrative support and it is essential that the safety auditor considers these issues at the appropriate time. Some additional administration issues which shall be investigated at safety audit concern:

- fire drill procedures;
- fire equipment maintenance procedures;
- fire certification procedures where appropriate;
- first-aid procedures;
- first-aid treatment room/recovery room procedures;
- lighting procedures;
- control of dust and fumes procedures;
- noise and vibration procedures;
- provision and issue of protective clothing procedures;
- storing and maintenance of protective clothing/safety equipment;
- manual handling procedures;
- transport procedures;
- welfare procedures;
- office/works cleaning procedures;
- hygiene procedures;
- safety library procedures;
- safety committee procedures.

This list is not exhaustive and you should develop it to suit your own working environment. However, all of these issues do require administrative procedures to be in place if they are to be efficiently and effectively operated in the workplace. Data about these matters should be systematically gathered throughout the year so that the safety auditor need only examine the data and observe a random selection of specific procedures (task analysis – see Chapter 4) in operation in order to satisfy himself that the administrative procedures are satisfactory (or not).

COMMUNICATION

Communication plays a central role in all administration for administration is communication. Likewise, management is communication. For example, if a revised safety procedure has to be implemented, everyone affected by it will have to be told about it. Therefore, we communicate to everyone the changes that have been made. Safety communication is not a secret art, it is a highly precise science. The way that we communicate can only be changed by careful planning. Some consultants may offer us instant solutions to our problems by handing us the magic spell....a formulaic prescription which will transform our safety communication. Safety communication is only improved by the application of general principles in an appropriate way in the specific context of the safety manager's environment.

Some traditional beliefs can create real barriers to changing the way we communicate, and to improving our personal effectiveness as managers and supervisors (see Chapter 9 of the Handbook of Safety Management (1992)). Safety communication is like each of these, and it is also like none of them. It is like each of them because it includes them; it is like none of them because it is a combination of all of them, and as such is more complex.

These issues involve:

- Assembling component elements: words and phrases, topics and themes.
- Sequencing actions, and pacing the performance.
- Setting goals and choosing the appropriate means.
- Noting feedback, and modifying plans and actions accordingly.
- Choosing actions from a repertoire.
- Adjusting to the opposition, and taking into account the resources available on your own side.

In order to identify the characteristics of good listeners, we find that:

- Listening is our primary communication ability. According to various studies, we spend about 70–80 per cent of the time we are awake in some kind of communication activity. Typically, for people in middle-class, white-collar occupations, of the time that we spend communicating, about:

  - 9 per cent of it is spent writing;
  - 16 per cent of it is spent reading;
  - 30 per cent of it is spent speaking;
  - 45 per cent of it is spent listening.

Although, in terms of time spent, listening appears to be the most important of our communication abilities, and writing the least, it is interesting to look at our educational system to see the priority that is given to the learning of the different communication skills which are:

	Amount of training	*Order of Skill Learning*
Writing	most	4th and last
Reading	next most	3rd
Speaking	next least	2nd
Listening	least	1st

The above table shows that, in the typical person's life, listening is the skill we learn first, whereas writing is typically the skill learnt last. The amount of formal training given to these skills probably reflects this ordering, with the skills we are least aware of, because they were learnt earliest, receiving the least attention. A rational training would be planned on a different basis, giving greatest priority to the skills that were most important.

Our listening habits are not the result of training, but rather the result of a lack of it. Because listening is the first communication skill we develop, and the least taught, we could assume that it is a well-practised skill which is normally done well. This is not borne out by the evidence available. Tests have shown that immediately after listening to a 10-minute oral presentation, the average listener has heard, understood and properly evaluated and retained approximately half of what was said. Within 48 hours that has dropped to half again, giving a final retention of no more than 25 per cent.

The value of good listening, and the problems created by poor listening are now well known. Recent surveys of safety managers have shown that listening, and listening-related skills, are cited as the most critical managerial skill, and the one in which training is most needed. For instance, Borman and his colleagues report that 'the most essential attribute of a good manager was the ability to listen'. In a study by Robert Likert (Harvard Business Review 1959) it was found that '95 per cent of foremen believed that they understood their subordinates' problems whereas only 34 per cent of their subordinates thought that their foremen understood them'.

It is known how important good listening is from our own experience of life. The problem is to identify the skills and techniques of good listeners.

There are three major reasons why good listening is valued in this way:

1. From the point of view of the listener, it is the source of information which is vital to both the successful completion of the task and the maintenance of the relationship.

2. From the point of view of the speaker, it demonstrates that they are valued by the other person. Bad listening devalues the other person.

3. The recognition of the importance of listening recognises the two-way nature of communication.

A number of studies have shown that safety managers are not very good at obtaining information from workers, and that they need the extra information that good listening would provide. It is estimated that some 40–60 per cent of the first accident analysis diagnoses made by safety managers are incorrect, and that this often arises because they fail to obtain the information they need, and that the worker can provide.

The best analogy here is, perhaps, of the accident investigator seeking clues in order to determine the cause of a chemical explosion. The solution emerges in the way that the clues fit together, not in the meaning of any one clue. What is important is that it is the pattern (the contributory factors) provided by the earlier clues which guides the accident investigator to seek further information, and to recognise its significance when found. If the accident investigator simply sits around waiting for all the forensic information to accumulate it is unlikely that any clues will be found at all.

However, when examining communications at safety audit it is necessary to take note of the time taken to effect any necessary changes and to note the way in which the message or messages were transmitted to the workforce. At the same time it will be necessary to record the method or methods by which the message(s) were communicated. Further information is given in Fig. 6.2.

TIME MANAGEMENT

The safety auditor will need to examine how key staff, such as managers and supervisors spend their time and to quantify the amount of time spent on safety-related duties. In order to assess time spent on various activities it is usual to arrange for a series of checks to be made. These usually involve the manager or supervisor recording the amount of time they spend dealing with safety-related matters. This data is useful to the safety auditor in the following ways:

- it allows the safety auditor to quantify the time spent on health and safety matters in the workplace;
- staff will be able to assess their own involvement in health and safety matters;
- staff will be able to quantify how much time they spend on the key task areas of their respective jobs.

There are many methods used to record individual time and some of these may have already been used in your own organisation. At safety audit it will be necessary to obtain time management data for examination. This is particularly necessary in known high risk locations.

INTERNAL AND PUBLIC RELATIONS

Industrial relations are an extremely important aspect of the safety management function. A good industrial relations procedure in the workplace will contribute to the overall accident reduction prevention mission and will improve communication, provide better job satisfaction and encourage a more active part in the organisational safety activities. Of course, industrial relations here do not just mean trade union involvement. Whilst trade unions do play a very important and necessary role in accident prevention it is also for management and supervisors to develop good working relations with their employees. This can simply be checked at safety audit by asking various people about other people in the group, team, section and/or department and asking a series of simple questions such as:

- Do you know the name of your supervisor's boss?
- Does he regularly visit, speak or write to you?
- Does he know what you do?
- Is he aware of your role?
- When did you last see the safety manager?
- What was the purpose of his visit?

Many questions can be devised in order to get a feel for the internal relations in an organisation and the above are given only as an example. In high risk locations this is a very necessary procedure to go through at safety audit.

In terms of public relations it is important that the organisation has a procedure for informing the general public about any high risk activities which are being undertaken and to involve them in matters which might affect the environment. An excellent example of good public relations is British Petroleum (BP at Wytch Farm in Dorset). Here BP actively involve the community, feel that they are a part of it and regularly keep the local population informed and involved in what they are doing.

Safety auditors will examine both internal relations and public relations and their procedures.

RECRUITMENT

Organisations are constantly involved in the procedure of attempting to select and recruit appropriate people for a job. For high risk jobs this procedure is important and should be subjected to safety audit. This process has become sophisticated and involves a considerable amount of time and effort particularly if effective recruitment and selection techniques are to be developed. A basic outline of the recruitment and selection process is discussed below but it must be stressed that each organisation will need to develop the process for its own ends. However, the fundamentals remain the same and the basic stages to consider in the recruitment and selection process are:

- Job analysis or description: have an understanding of the job to be filled.
- Job specification: understand the knowledge, skills and aptitudes required to do the job.

- Recruitment media: decide where suitable applicants are to be found.
- Advertising: decide how people can be persuaded to apply.
- Job administration: choose the means of finding whether the applicants have the required knowledge, skills and aptitudes. Consider methods of application such as application forms, interviews, references, selection tests, etc.
- Rating schemes: make a choice of which method to be used for scoring applicants.
- Induction: introduce the successful candidate to the job and to the organisation.
- Evaluation: assess the recruitment and selection process.

The features of the recruitment and selection process outlined above were dicussed in Chapter 5. In situations involving unacceptably high accident rates or dangerous occurrence rates or in the case of major disasters it would be appropriate to include the recruitment procedures, job specifications and job descriptions as an integral part of the safety auditing procedure.

SAFETY TRAINING

Evaluation of safety training

An evaluation of safety training, particularly at management level, is an extremely difficult process and it is for this reason that many organisations balk at the thought of attempting to carry it out. However, if the training provided is to be improved, some evaluation is necessary to determine where the shortcomings lie. There are four levels of evaluation which can be considered:

1. Reaction level.
2. Immediate outcome level.
3. Intermediate level.
4. Ultimate level.

These are now discussed in more detail.

Reaction level
Here, the safety training is evaluated on a subjective basis by the trainee immediately after the safety training has been completed. This is a very superficial evaluation but can contribute to the overall evaluation process. The sort of questions that might be asked would concern:

- The presentation of the information.
- The presentation of material.
- The relevance to the trainee.
- The usefulness of the training.

Immediate outcome level
This level of evaluation involves some objective testing of the trainee in terms of skills learnt or safety knowledge acquired. The trainee can be tested before the training takes place and again after it has been completed and the difference can be quantified.

Intermediate level
Here the evaluation is made some time after the safety training and focuses on changes that have taken place in the trainee's actual job performance. This kind of evaluation is particularly concerned with the transfer of safety training to the job. This is a critical area as often the trainee may be happy with the training provided but be unable to put it into practice. The problem at this level, is that there are many factors which might have affected the trainee's job performance other than the training itself. To relate any improvement in job performance to the training (to the exclusion of all other factors), is a difficult process.

Ultimate level
The ultimate objective for safety training should be to improve overall organisation effectiveness which will include its safety record. It can be argued that true evaluation of safety training only takes place where the safety training is linked to some improvement in organisational performance. If safety training is to be worthwhile, the attempt must be made to ensure that it does lead to an improvement in organisational effectiveness otherwise it will become difficult to justify.

Safety training should be an important activity in any organisation and it does not necessarily always need to be course based. Safety training forms a part of the overall safety mix and should not therefore be used in isolation but should be linked to some other aspect of the mix. Managers should understand the importance of developing their subordinates' safety awareness, and thus realise that safety training, in either a formal or informal sense is an important part of their role and cannot be totally delegated to a training department or ignored.

The value of induction training

Safety training is usually carried out by an organisation in order to help recruits to overcome their sense of strangeness, secure their acceptance by existing employees and develop in them a sense of belonging. A large amount of labour turnover occurs during the early weeks of employment usually because no effort has been made to enable the newcomer to feel welcome. As a result of this he becomes unhappy and leaves to find alternative employment. This sort of training is an opportunity to introduce the new employee to the organisation's purpose and to discuss its safety policies, procedures, practices and programmes thus seeking to establish the correct link between each individual, his work and activities external to the organisation. In addition there is an opportunity to:

- Outline the new employee's place in the company.
- Outline the relationship between the employee's work and the finished product.
- Outline the relationship between the employee's company, industry and community.
- Outline the company's positive attitude to health and safety.
- Outline how the new employee can put ideas and points of view to management.

For smaller firms, the induction course is an informal affair and carried out normally by the recruit's immediate superior. It will be necessary to examine safety training syllabi and to observe the ways in which safety training is provided.

Training programmes in safety subjects should be conducted as often as the number of new recruits warrants and should be organised by the personnel department. It is important that senior executives are involved as this gives a good impression to the recruit and also adds prestige to the occasion. The safety manager might be responsible for conducting a tour of the workplace in order to explain:

- How each job fits safely into the flow of production.
- Departmental rules, procedures, practices, discipline, etc.
- Accident prevention procedures.
- Protective clothing policy.
- Action in an emergency.
- Hygiene rules and regulations.
- General safety and health policy.
- How to contribute to the well being of others, trade union membership and knowledge of members of the safety committee.
- Safety training policy.

If these points are to be covered in written form it is important that the information is well presented, readable and kept up to date. Occasionally tenth generation photocopies are used for this purpose and it is important to ensure that sufficient copies of the original are available for recruits. It is worth holding a follow up meeting at a later stage for these recruits in order to evaluate the induction process.

Safety training is examined in detail at safety audit and will cover such items as frequency, instructor competency, delivery, appropriateness of syllabus, suitability of resources and teaching premises, content, methodology and safety training management procedures.

SUPERVISION

This is an extremely important part of the safety management process and good supervision is an essential part of the accident reduction/accident prevention strategy employed in organisations. It is important that supervisors are selected by using recognised human resource methods, that they understand their roles, tasks and objectives and are trained regularly in the safety areas of their jobs. It is also important to include in the safety audit the supervising of the supervisors and the same criteria will apply.

Good supervisors will be able to demonstrate good leadership qualities and this should be one of the leading human resource criteria applied to the selection process. It is important that supervisors involved in high risk activities are audited thoroughly including an examination of the selection processes. There would also be a requirement to investigate these criteria in cases where the accident rates were unacceptably high or where too many dangerous occurrences were taking place.

A note should be made of all training courses undertaken by the supervisor since the last safety audit and special note made of safety training courses attended. Courses completed by attendance only should be separated from those which required some form of assessment.

DISCIPLINE

It is an essential part of the safety mix that sound rules are in force in order to contribute to the company accident reduction programme. If we were to have a football match without a referee it would not be long before individuals made up their own rules and total confusion would then ensue. It is necessary in an organisation to have disciplinary procedures in place which should be communicated to each employee in a language that can easily be understood. There are usual rules in place regarding such matters as:

- smoking;
- wearing of safety clothing;
- wearing of hats;
- wearing of safety glasses;
- wearing of gloves and shoes.

The list could vary from industry to industry but the point is that at safety audit an individual list should be drawn up so that the disciplinary procedures and records can be properly investigated. Other companies would include rules about the use of equipment, authorisation of users and even skylarking.

An example of a safety procedures audit checklist is given in Fig. 6.2.

Activity	Satisfactory	Not satisfactory	Improvement since last safety audit
Administrative structure Checklist of relevant regulations Are administrative safety procedures efficient and effective (list shortfalls and note good practice)? List individual roles Have all administrative performance standards been met (list those which have not)? Test administrative hierarchical structure for safety performance achievements Assess administrative section/dept. safety performance as a group			
Communications Examine language used in all safety publications, literature and letters since the last safety audit for ease of reading by all relevant groups Examine frequency of safety communications by topic and relevance. Has all new information been passed on and in what time scale?			
Time management Examine overtime records List all key workers who work more than 45 hours per week Examine performance of these individuals			
Internal & Public Relations Examine all internal publicity exercises and exhibitions aimed at improving internal safety performance since the last safety audit. List the key task areas and test these for achieving all aims and objectives. A similar exercise will be undertaken to assess external publicity and other public relations exercises			
Recruitment List all personnel who have joined the company since the last safety audit and examine all safety literature, training and equipment provided for their use. Examine personnel procedures for safety promotion in their use Examine personnel procedures for safety promotion in their recruitment programmes Examine samples of newly-recruited key personnel against accident records for an assessment of safety performance			

Figure 6.2 A safety procedures safety audit summary checklist

Activity	Satisfactory	Not satisfactory	Improvement since last safety audit
Safety training List all safety training courses held since the last safety audit Examine syllabus for relevance and test appropriateness Examine safety training frequency and refresher courses in relation to high risk activities and known accident situation Check classroom facilities and safety training equipment/literature Examine quality of instructors/lecturers Examine safety training curriculum			
Supervision List all high risk activities by department and at section level Examine supervision available and qualifications and experience of each supervisor for safety competence and ability Examine attendance records of all key supervisors and examine their safety training records. Examine all accident records relevant to each supervisor and their section accident and dangerous occurrence records Examine individual supervisor's overtime records and note all those who work 45 hours or more per week.			
Discipline Examine all disciplinary records since the last safety audit by section, department and by individual supervisor Examine all dangerous occurrence records and action taken Examine all accident records and check all action taken as a result Examine disciplinary procedures at supervisory, senior management and administrative levels Examine appropriateness of the disciplinary decisions taken against the overall company philosophy Examine the involvement of trades unions and safety representatives in disciplinary disputes and their contributions made to accident reduction and prevention strategies			

Figure 6.2 cont. A safety procedures safety audit summary checklist

7 AUDITING SAFETY PRACTICES

In this chapter we will look at the following:

- Costing and evaluations of accidents
- Accident investigation
- Data collection
- Medical examinations
- Welfare
- Hazard and risk assessment
- Accident analysis
- Equipment and system inspections
- HSE Codes of practice
- Professional Codes of practice

COSTING AND EVALUATIONS OF ACCIDENTS

There are many different methodologies used by different companies to calculate the cost of accidents in relation to their activities. Some have been discussed in the Handbook of Safety Management op.cit and others have been put forward by other researchers and safety cost accountants. For example, Saunders and Ordoqui (1992) produced a paper entitled *'Calculating the cost of accidents and the benefits of safety programmes in developing countries'* which not only offers a methodology for calculating accidents and the benefits of safety programmes but also provides some guidelines on the level and type of information which should be considered when costing accidents including some information on calculating the worth of the employee to the organisation. In addition, the HSE have considered this area to be important and have undertaken a number of trials in different kinds of industries. They have developed a methodology which is worthy of some discussion here.

In the methodology put forward by the HSE they consider the real cost of accidents as having three distinct factors:

- direct or insured costs;
- indirect costs;
- intangible costs.

For example, direct costs would include those costs which are insurable such as:

- injury costs;
- illness costs;
- damage costs.

whilst indirect costs (which are uninsured costs) would include items such as:

- product or material damage;
- building damage;
- plant damage;
- equipment damage;
- legal costs;
- expenditure of emergency supplies;
- production delays;
- equipment down time;
- site clearance and emergency repairs;
- overtime working;
- accident investigation time;
- diverted supervision time;
- enquiry time;
- administrative costs.

Intangible costs might include:

- damage to company image;
- loss of business;
- lowering of morale;
- poor recruitment in the future.

Safety audits must be able to compare accident costs with previous ones and it must also be possible to compare economic rates of return. If your company does not know what accidents costs were incurred last year or since the last safety audit then it is important that one of the methodologies approved by your cost accountant is implemented as soon as possible.

When examining accident-costing methodologies for safety audit purposes it is necessary to try and quantify the levels of under-reporting of accidents which can occur so that these may be considered in the economic analysis. The level and type of question which should be answered in the accident-costing process is summarised in Fig. 7.1.

It is important that your company cost accountant agrees to the content of the questions listed in this Figure.

	Accident 1	Accident 2	Accident 3	Accident 4 etc.
How many other workers (not injured) lost time because they were talking, watching or helping with the accident?				
How much time did they lose?				
How many other workers lost time due to damaged equipment?				
How much time did they lose?				
What is the cost of equipment repair or damage?				
How much time did the injured worker(s) lose on the day(s) for which he was paid?				
What overtime was paid in order to restore lost production?				
How much supervision time and costs were involved?				
How much accident investigation time and costs were involved?				
How many days' absence have been incurred as a result of the accident?				
What costs have been incurred by this lost time?				
What are the medical costs involved?				
Additional costs (please list)?				

Figure 7.1 Some accident-costing questions to ask at safety audit

ACCIDENT INVESTIGATION

Studies have found that some organisations keep no accident or dangerous occurrence records other than those provided for within the requirements of Form F2508. Many do not computerise the data but maintain expensive and cumbersome manual systems. Accident data collection by some companies is seen merely as meeting the requirements of statutory obligations and few exploit the value of such costly information as a management tool.

Other researchers have found that people are more likely to forget facts and may not be able to recall up to half of what they originally perceived when questioned as little as 24 hours after the event. Gathering good reliable information must apply to the role of the personnel manager or indeed any manager and not necessarily for use in disciplinary matters. Whilst the casualty might be excused from recalling certain information concerning an accident because of shock, it is important that procedures are in place which ensure that witnesses are spoken to within 24 hours whenever possible, and that the interview is structured correctly to eliminate some of the problems outlined above. Safety auditors must seek out this information and should be able to quantify the delays taken to take statements or to state the reasons for any delays encountered.

It is now known that information published by the HSE is not sufficient in itself to assist the supervisor or manager in programme planning. It lacks sufficient detail. Contributory factors are not identified in sufficient depth to assist with the identification of trends. Once contributory factors are identified then appropriate remedial measures can be introduced using the safety mix i.e. by enforcement, education, engineering and environmental strategies or a mix of some or all four. Making something illegal will not prevent it from happening and there will always be a need to combine enforcement with one or more of the other factors or elements within the safety mix.

Because of poor information gathering, a large number of safety management decisions are based upon opinion. In order to get good information for safety audit purposes in all accidents one seeks to answer six basic categories of question. These are:

1. Who?
- Who was involved in the accident?
- Who was the line manager responsible for safety?
- Who reported the accident?
- Who was called to respond to the accident?
- Who was notified?
- Who should have been notified?
- Who was responsible?

2. When?
- When did the accident occur?
- When were people aware that an accident had occurred?
- When did help arrive at the scene of the accident?

3. Why?
- Why did the accident happen?
- Why were safety practices not applied?
- Why did safety procedures fail to work?

4. What?
- What actually happened?
- What were the losses incurred?
- What injuries were sustained?
- What could have been done to avoid the occurrence?

5. Where?
- Where did the accident occur?
- Where was the safety officer or line manager at the time of the accident?

6. How?
- How did the event occur? For example, rapidly, slowly, without warning?
- How could safety procedures and practices have been improved?
- How does the organisation learn from the accident occurrence?

It is known that human errors and failings contribute with other factors to the vast majority of accidents and dangerous occurrences. Failing to cope with circumstances leading up to and prevailing at the time of the accident will be evident in most situations. It is important that all accident investigations consider human error and it is therefore vital to identify prime factors involving human behaviour as a major part of the accident investigation, together with human reaction to unsafe conditions within the workplace which should also be included. The safety auditor should be in a position to identify human behaviour issues from the accident data and to identify company attempts to prevent or reduce the incidence of these.

There are two primary objectives of accident investigation at safety audit and these are:

Accident reduction
Obtaining sufficient data to facilitate the systematic reduction in the type and severity of accidents in the workplace. These will fall into the four categories identified above as the safety mix and will provide information of sufficient quality to balance the mix to best effect.

Accident prevention
This relates to the application of safety principles in new design and technology whether it is in the area of automation or improvements to the style and type of manufacturing methodology employed in the workplace.

Both of these strategies have to be based upon recognised opportunities which are available whether to the designer, engineer, enforcer or general safety practitioner for influencing and preventing all forms of accidents from happening. Both strategies require the co-operation of everyone to succeed and must not be left to one person to solve alone. Only in this way can effective strategies be employed.

It is important that the safety representative is involved in all accident investigations and that he is fully aware of the aims and objectives of the data collection process. It must be appreciated that in certain circumstances accurate data collection can be hampered by the 'high status' phenomenon and care must be exercised so that the investigation is not seen as a purely disciplinary or enforcement procedure. Such investigations may assist in the formulation of rules and safety procedures but should not be seen as a means to an end. Obvious breaches of established procedures, however, must be dealt with in the interests of everyone concerned. Such weaknesses must be looked for within the safety audit process.

The most important part of the manager's operational daily safety plan is the systematic collection of accident and dangerous occurrence data for analysis purposes. It should be remembered that many dangerous occurrences happen in reality but never get reported. This may be due to a feeling of failure by the person concerned, and threats of disciplinary action for breaches of safety codes and practice will not encourage the reporting of such incidents. Some organisations have set up confidential telephone lines in an attempt to combat this problem. The safety audit would need to check for these schemes. Some firms can also show an increase in the reporting of dangerous occurrences but admit that they may still not be aware of all such instances. Safety management, therefore, must have a structured and systematic approach towards the gathering of dangerous occurrence data. Whilst the confidential telephone idea is a sound approach to the problem it should not be relied upon as the only method of gathering information. Confidential information forms can also be used and these can be structured in such a way as to collect contributory factors required in the subsequent analysis. At the same time, the safety manager should have sufficient standing within the organisational structure to be able to interview any member of staff involved in a dangerous occurrence in total confidence. This method is also an important way of obtaining data. It is very likely that an employee involved in such an incident may feel worried by it. He may even feel fright or shock. When a dangerous occurrence takes place it is usually regarded as a good learning vehicle but this practice must not be encouraged! At the same time, if a lesson has been learned by both employee and employer what need is there of disciplinary action?

Having given thought to primary data obtained for safety audit purposes, information should be supplemented wherever possible with other data published by government departments including the HSE, the Department of Social Security, the Government Statistical Office and the RoSPA equivalent abroad. This gives the safety auditor an opportunity to put the primary data into perspective. The HSE in the United Kingdom annually publishes accident and disease types data but does not publish those contributory factors evident in each accident, disease or dangerous occurrence. In fact, the current official data collection form F2508 does not seek such factors contributing to these incidents. It will be necessary, therefore, to collect these locally if the safety mix is to be planned, implemented, monitored and evaluated efficiently and effectively. To provide for the simple evaluation of data collected, it should be gathered in a standardised form. Bearing in mind that the results may be placed upon a computer, thought needs to be given to the coding of the results. To facilitate this, the data collection process can be broken down into the following broad headings:

1. Accident details.
2. Casualty information.
3. Details of attendant circumstances.

For administrative purposes, provision will need to be made for the identification of accidents individually and for general recording purposes. Data required by the enforcement agencies should form a part of this process.

Safety audit of the accident details

In general terms, this section must describe what happened, when the incident occurred and the location of the accident site. Within this section it is important that details of all equipment involved at the time of the accident are also recorded. At the same time, accident damage is noted both to equipment and surrounding environment together with details of independent witnesses. It is also important at this stage to note any procedures adopted and whether appropriate safety precautions were taken and that issued safety clothing was worn. Whether procedures were not carried out or clothing not worn is not at issue here. Questions relating to why certain factors exist are dealt with under attendant circumstances and are deemed to be contributory to the accident or occurrence.

Safety audit of the casualty information

Here it is necessary to record details of who was involved in the incident and the levels and type of injuries received. Generally, it is important to record those killed or injured. Few accident investigators support the classification of injury types at this level. In an accident a person is either killed or injured. A non-injury may be deemed a dangerous occurrence. It is difficult to identify those separate factors which contribute to a serious injury rather than a minor injury or what makes an incident a minor one rather than a major one. For example, in two identical instances two people fall from a roof 20 feet in the air. One lands on his head and is seriously injured where the other lands on his feet and suffers minor lacerations. In this case, one employee was off work for two months whilst the other returned to work the following day. Where scientific research is involved there are accident injury scales for the classification of disease and injuries. It is argued that for the operational management of effective safety procedures and practices, these are unnecessary. It must be remembered that the aim of the exercise is to identify contributory factors present when an accident occurs. The sole purpose is to identify these and take remedial action. The fact that an employee was injured in varying degrees does not significantly contribute to this part of the exercise. It is for this reason that personnel managers need only classify injuries simply at this stage.

Safety audit of the attendant circumstances

This part of the procedure considers those factors which have contributed to the accident or dangerous occurrence. It will include details of the accident site and its immediate surroundings and will cater for factors which led up to and immediately preceded the accident or occurrence. Where a non-injury incident is involved it may only be necessary for a self-certification exercise by the person involved to be carried out. It is recommended that in these circumstances the casualty details be suitably modified to ensure anonymity.

Care should be taken that all questions that can be regarded as subjective are completed with care. For example, if an accident occurs outside then it might be necessary to explain the meaning of 'high wind', 'rain/drizzle', 'mist/fog' etc. in any instructions or in training given. These factors often appear in outside-type incidents. Indoor-type accidents can suffer a similar language interpretation if adequate care is not exercised. For example, the difference between bright light, light and darkness will need care as would matters referring to ventilation such as airy, breezy, windy, gusting, etc. In a recent case during a warm sunny spell, the workforce opened windows and doors to let in a cool breeze. The breeze was rather strong and blew over several items of equipment stored in the work area. These fell over and injured a worker's foot whilst another had dust blown into her eyes thus causing the worker to catch her fingers in a machine. Here, the contributory factors are identified as the weather, the open windows and doors, the windy, blustery conditions, the loose materials capable of being blown over, and the dust. The casualty rate could have been affected by the removal of any one (or more) of these factors. Similarly, a person slipped and fell in a car park because it rained for the first time in eight weeks and the rainwater mixed with oil and rubber caused the car park surface to become very slippery. The employee was late for work and in his haste to run into the office to avoid getting wet slipped and broke a leg.

Where man fails to cope with his environment, most data collected for accident analysis purposes can be standardised. There can, of course, be allowances made for those occasional variations which do exist between occupations and industries.

DATA COLLECTION AT SAFETY AUDIT

It is necessary to examine all aspects of the accident investigation process including the method by which accident data is collected and used for decision-making purposes. It is necessary systematically to obtain information relating to accidents which enables the management to identify contributory factors. One recognised form of gathering this information is by using a survey.

The purpose of surveys is to obtain information about accidents. Obtaining this information is never free and can be expensive in terms of time and money. Before collecting any sort of data, thought must be given about how relevant it is to the needs of the situation, how it can be analysed usefully and how it can be obtained. The following is a simple description of a very complex subject and further reading references are given at the end of the book.

Factors which undoubtedly need consideration when choosing the various sources of data are:

- the worth of the data;
- the ability to cross check key information;
- time needed to collect;
- cost of collection;
- accessibility (particularly internal records);
- political sensitivity.

Cost benefit aspects should be considered in making the choice. Clearly, subjective judgement will play an important role in this choice as the implication of various data decisions may be unclear.

There is also a great need to assess the ability to analyse data prior to collecting it. It is not uncommon to discover, after great effort and expenditure of time, that the following are true:

- There is too much data for manual analysis.
- The form of the data collected is not united with the form of analysis to be applied.

Some common types of analysis are:

1. Significance testing of means and variances.
2. Goodness to fit (X^2) tests.
3. Testing the correlation between variables.
4. Establishing a regression line.
5. Grouping data in various ways.
6. Preparing histograms, pie-charts, graphs, etc.

Sources of primary and secondary information

The establishment of a sound and credible information base is essential for any project evaluation. Common failings in many safety audit investigations are:

- A failure to identify potential sources (i.e. insufficient initial research).
- A failure to use sources (being deterred by effort, time, cost or political antagonism).
- Poor choice of sources (using subjective data where objective data is possible; failing to build in validation processes; failing to identify the sensitivity of the end decision to data accuracy).

A simple classification of data sources is:

1. Primary data – this is the information which originates from your investigation:
   (a) what people/systems do (observation and experimentation);
   (b) what people say (questioning and expert opinion).

2. Secondary data – this information does not originate from the investigation.
   (a) what people have done/said (internal records);
   (b) past performance of systems (external published sources).

Primary data

The fact of gathering data for specific needs often forces greater thought about:

1. What is trying to be done?
2. What contribution will the piece of data make?

Gathering data in this way often provides the opportunity for personal contact, obtaining specific pieces of feedback or just developing thought in order to provide a greater insight into the nature of the problem. Almost without exception, some cross checks are needed of secondary data via primary data and vice versa.

Observation

These methods use some kind of recorder, either human or mechanical (e.g. video cameras, or the human eye, etc.). It is the aim of such methods to avoid any abnormal behaviour of the system due to the observation. This is far from easy; frequently people have a tendency to settle to 'normal'. Examples of such an approach are:

- Activity sampling surveys (working on a new piece of equipment to record what happens).
- Filming a specific activity in a department.

- Traffic flow counters.
- Visual checks on floor utilisation.
- Safety studies as a part of the safety audit.

Experimentation

This would be used to examine differences from the previous safety audit. The aim is to identify the difference in the outputs of a changed system with those prior to, or in, an unchanged situation. The content of the use is in the testing of a cause-effect model which has resulted from assessment of the current situation. Examples of this might be:

- testing new equipment;
- testing new procedures;
- testing safety ideas.

Questioning

This is the most common source of primary data involving the use of interviews and/or questionnaires. Asking questions is a good method of establishing and cross checking causal models and is the only way to get at behavioural problems in an organisation. Considerable skill in the use of interviews and questioning techniques is required if data is to be obtained which is worth analysing as a basis for useful information. The ease with which this source can be misleading or inaccurate is evidenced by the disastrous results of schemes based almost entirely upon data from such sources. The cost and time attached to the use of this source is influenced by the geographic dispersion of the people involved, the number of people involved and the technique chosen. These issues are discussed later.

Expert opinion

There are certain situations where the processing of data to form information is best done by a particular person. Examples might be:

- Evidence from a scientist specialising in a particular field relevant to the study.
- Production controller estimating delivery of new equipment.
- Union representative on reactions of his members to a new procedure.

The reasons for accepting expert opinion in the main fall into the following three categories:

1. The rules for processing the data to form information are not available outside the mind of the individual.

2. The data cannot fully be established as it is in the subconscious as well as the conscious mind.

3. The worth of extracting the data, etc., is not offset by the benefits of greater accuracy (often a general indication is all that would be required).

Secondary data

Such data tends to be easily accessible and often provides a good starting point for an investigation. Desk work based upon such data can be helpful in the following:

- Defining terms, gaining understanding in new fields of knowledge.
- Widening view points on relevant issues.
- Establishing the current 'state of the art' in specific areas.
- Identifying the areas where primary data will contribute.
- Formulating possible approaches to resolving the problem.
- Obtaining some understanding of the environment under investigation.
- Providing a theoretical background to the study.

Internal records
The value of internal records for safety audit purposes has already been discussed and most organisations keep many internal records, some formally, some informally; these records can be of great value in project work once located and released. Skills in interviewing and gaining personal acceptance are often a prerequisite to this. As a base, internal records can reveal accounts, costings, labour and some accident statistics.

External published data
A large volume of current, published data is accessible, some of which is almost bound to be relevant to a specific project. Examine whether these have been obtained from professional organisations, trade associations, libraries and colleges of higher education. It is usual for any article to be obtainable within 14 days.

Since most problem situations are by no means unique, it is highly likely that there will be some comment in the literature relevant to the problem being investigated. This being so, it is surprising that the relevance of external published data is underestimated. However, it is important that when using external data sources the context of any findings is assessed for validity prior to local use.

Comparative studies

The result of collecting primary and secondary data will provide a base for formulating a hypothesis about future behaviour of systems or people which will be significant in the recommendations of any change. The use of comparison and analogy can often provide further insight and affect the confidence in a model. The utility of this method lies in the suitability of the comparison chosen.

Comparative studies can offer some of the following:

- a cross check on how to solve a problem;
- an indication of similarities and differences in approach;
- identification of the results and associated side effects to compare with predictions.

Accident data collection

Decisions will always be influenced by a person's perception of those elements of the environment, internal or external to the organisation, which are seen as relevant. Thus, a greater understanding of issues and their relevance to a particular decision along with a more realistic picture of the environment will lead to a better basis for decision.

Information is required for many reasons, some of these are to:

- clarify the request of the client;
- identify issues relevant to the problem or the decision;
- enable projections to be made of future behaviour of people and systems;
- establish the current situation and what has led up to it;
- formulate an approach for resolving the problem;
- check personal ideas against those of others.

There are eight steps involved in establishing a data or information base. These are:

1. An identification of the data requirements.
2. An identification of potential sources of data.
3. A selection of sources to be used.
4. A selection of the methods to be used for collecting data.
5. Planning the method of data collection and the analysis which will follow.
6. Collecting the data.
7. Analysing the data.
8. Presenting the information in a clear and understandable form.

Data often exists in a form which has little value as an aid to specific decision making. In converting data into a meaningful form of information, one or more of the following should be considered:

- aggregation or disaggregation of the data;
- regrouping and/or ordering of the data;
- testing relationships between various types of data;
- trying various methods of presenting the data.

Medical data since the last safety audit	Number
Total number of company patients seen since last safety audit? Total number requiring treatment? Total sick absence authorised (three days or more)?	
The following is a sample of the information which should be available but safety managers should arrange for their own specific needs in discussions with their medical practitioners. This list, therefore, is merely an example. Number of cuts treated Number of abrasions Number of bruises Number of fractures Number of burns	
Number of complaints concerning: Noise Vibration Temperature Ventilation Lighting Dust and fumes Radiation Stress Alcohol	
Dangerous and hazardous substances	

Those organisations who do not employ a medical practitioner should obtain the necessary information from the first-aid registers and qualified first-aid personnel. RIDDOR requirements may be available from the Personnel Department but the safety auditor should request this information from whoever is responsible for it.

Figure 7.2 An example of data a safety audit would require from the company medical practitioner

It is often only by trying different approaches to, and combinations of the above, that insight and ultimate understanding can be achieved within the safety audit.

EXAMINE MEDICAL RECORDS

The safety auditor should take care when examining medical records as these are protected by medical confidentiality. Most company doctors will provide a general statement of records held in surgery but it is unreasonable to expect details of the identity of individuals in the safety audit. Most company medical practitioners will provide details as illustrated in Fig. 7.2 for safety auditing purposes.

An additional part of the medical enquiry stage would include an examination of the sick records held at personnel, department or section levels. In particular information required to be reported under RIDDOR should also be available as part of the safety audit. It is usual to also obtain information from supervisors and managers as to the action they have taken as a result of the medical enquiry information gathered from the last safety audit or as part of the general safety management process.

WELFARE

Welfare covers not only washing facilities as discussed in Chapter 5 but also includes the following issues. All sanitary conveniences, provided by an employer must be kept clean and there must be provision for lighting, ventilation and privacy. Where persons are employed of different sexes then proper and separate accommodation must be provided. There are rules concerning the number of conveniences outlined in the Sanitary Accommodation Regulations, 1938 as amended in 1974. They are summarised in Fig. 7.3. These must be a part of the safety auditing process.

Alcohol and drugs

Alcohol has been identified as a contributory factor in a number of accidents in the workplace. It is important that workers are made aware of the dangers of drinking either before going to work or during it. Workers who drink heavily the previous night may still be over the legal limit to drive a car the following day. It is unlikely that anyone unfit to drive a vehicle will be fit to operate machinery or equipment at work. An organisation must take steps to include in its policy on health and safety its stance over the use of alcohol and medicines. In cases of drinking alcohol personnel managers must consider:

- action to be taken in a one-off situation;
- action with a persistent offender.

Minimum number of conveniences		Actual
	Toilets	Toilets
For every 25 female employees For every 25 male employees	1 1	
Where the number of males exceeds 100 and sufficient urinal facilities are also provided	4	
For every 40 employees over 100	1	
Where the number of males exceeds 500 then for every 60	1	

In counting the number of employees, any odd numbers less than 25 or 40 are regarded as 25 or 40.

When examining sanitary conveniences it is also necessary to examine the condition and cleanliness of these and the provision of appropriate toilet paper and soap. Adequate sinks and basins should be provided and should not be cracked. All taps should be serviceable. Under no circumstances should kitchen activities (washing cups etc.) be carried out in hand wash basins.

Figure 7.3 A summary of the Sanitary Accomodation Regulations, 1938 (amended) 1974. The safety audit will examine the actual sanitary provision against the regulations

In the case of the latter it may be possible to offer counselling advice whereas the former might be easily dealt with under a disciplinary investigation depending upon the circumstances. A similar system would be required where workers were found to be dependent on unprescribed drugs.

Workers who are taking prescribed drugs which medical practitioners have identified as being capable of clouding judgement or affecting manual dexterity should be afforded some protection and light duties may be a possible solution in the short term. Workers should be encouraged to ask their general medical practitioners if any medicines that they have been prescribed will either affect their ability as a road user or as a worker.

Stress management

Being able to identify stress in workers must be encouraged. Being able to do something about it is the management function. Stress has long been identified as a contributor to accidents and it is therefore important that it is identified and dealt with. Stress can affect people in different ways. Stress can be eased by relaxing and building confidence, therefore the following are important:

- safety training;
- safety education;
- task analysis and the identification of areas of risk;
- bonus schemes and/or time management;
- inter-personal relationships;
- assertiveness;
- career and/or life planning;
- counselling;
- self-awareness;
- negotiation;
- team building;
- performance management.

However, stress management is best considered within the job profile and recruitment stages and these issues are developed further in other books in this series.

There have been attempts to quantify levels of stress and one such example is given in Fig. 7.4 referred to as the Holmes-Rahe scale of life change units (LCUs).

Cleaning

There is a requirement under the HASAWA for an employer to ensure a safe and healthy working environment. This will require premises to be kept clean and tidy. Most organisations sub-contract this responsibility to office cleaning companies or undertake the task themselves. Office or factory cleaning usually takes place after everyone has gone home but as the name implies cleaning also includes good housekeeping whilst actually at work. Poor housekeeping is responsible for more than half of all work accidents. People should not:

- block or litter walkways/gangways with raw materials or other 'essential' job items;
- carelessly discard food or liquids either over floors or equipment;
- throw or discard items around the work area which people can trip over;
- hang clothing on equipment not designed for the purpose;
- cover up important notices or signs;
- use waste containers incorrectly;
- leave rest rooms and/or kitchen areas littered with motorcycle helmets and other equipment for people to fall over.

Event	LCUs
Death of a spouse	100
Marital separation	65
Death of a close family relative	63
Personal injury or illness	53
Marriage	50
Loss of job	47
Marital reconciliation	45
Retirement	45
Change in health of a family member	44
Wife's pregnancy	40
Sex difficulties	39
Gain in family membership	39
Change in financial status	38
Death of a close friend	37
Job change	36
Argument with spouse	35
Taking out mortgage	31
Mortgage foreclosure	30
Change in work responsibilities	30
Sibling leaving home	29
Trouble with in-laws	29
Personal achievement	29
Spouse beginning/stopping work	29
Change in personal habits	24
Trouble with business superior	23
Change in work hours or conditions	20
Change in residence	20
Change in schools	20
Change in recreation	19
Change in social activities	18
Taking out a personal loan	17
Change in sleeping habits	16
Change in family social arrangements	15
Change in eating habits	15
Vacation	13
Minor violation of the law	11

It is important in accident investigation and at safety audit to examine the presence of any of these factors as contributors to the accidents being audited.

Figure 7.4 Holmes-Rahe scale of life-change units

Hygiene

Employers should set out rules for the consumption of food and drink on the premises. Also, those organisations providing works canteens will be subject not only to the HASAWA but also the Food Act, 1984 and those regulations made under it. It provides for:

- *Injurious foods* in that it is an offence for any person to add any substance to food, use any substance as an ingredient in the preparation of food or subject food to any other process or treatment so as to render the food injurious to health, with the intent that the food be sold for human consumption in that state.

- *Protection for purchasers* whereby it is an offence to sell to the prejudice of the purchaser any food which is not of the nature, substance or quality demanded by the purchaser.

It is also an offence to sell food which is unfit for human consumption under section 8(1) of the Food Act, 1984.

Where food is kept or prepared, care should be taken to ensure that sink units and refrigerators are well maintained. Personal hygiene should be monitored and provision made for washing and cleaning. Implements used for eating or drinking should not be washed in toilet areas and cracked sinks should be replaced as soon as possible.

HAZARD AND RISK ASSESSMENT

The identification of hazards is an essential first step in risk control and information is required and reference should be made to:

- all legislation and supporting codes of practice;
- appropriate HSE guidelines;
- product information;
- BSI or ISO standards criteria;
- industry or trade association guidelines;
- local knowledge and experience;
- accident data;
- expert advice and, as a last resort, 'opinion'.

The safety auditor should conduct a critical appraisal of all activities and take note of all activities which are hazardous to employees and those other people affected by the activities of the company. It is important that details of role, task and performance standard data are available to the auditor in order to facilitate the hazard identification process. Employee and safety representative input is also a useful contribution. In some industries, hazard and operability studies are carried out and these are referred to as HAZOPS. Where such studies are used the safety auditor will also require access to this information. Additionally, some companies use a technique referred to as 'fault tree analysis' and in cases where this has been used the safety auditor would require sight of the information too.

Assessing risk, on the other hand, is required so that appropriate control measures can be assessed by the safety auditor. Determining the importance of risk involves an assessment of the severity of the hazard and the probability of an occurrence. Here, accident data is the most useful tool in risk assessment. Such data should give an insight into whether there has been an 'occurrence' since the last safety audit. Known risks must receive a high priority within the decision-making process. In some industries, a qualitative assessment might be necessary within the framework of legal requirements. The Control of Substances Hazardous to Health (COSHH) Regulations, 1988 and their accompanying approved codes of practice would be a good example of this. The safety audit should examine these areas closely.

There is no general formula for assessing risk in practice, there are a number of techniques which have been developed to assist the safety auditor. These, however, must be separated from the detailed risk assessments which are required to establish appropriate levels of risk control necessary to meet legal requirements. There is also a danger of apportioning numbers which might give management a false sense of security. The techniques suggested here provide only a means of ranking hazards and risks but great care should be taken when calculating the probability of an 'event' occurring. There is always the chance that it will occur!

Hazard estimation should use the same criteria used in accident investigation. There are four categories which should be rated as follows:

- Fatal.
- Serious.
- Slight.
- Damage only.

It is appreciated by the safety auditor that injury and/or damage might not occur in every case and it is normal in practice that the likelihood of injury will be affected by the organisation of the work as identified through task analysis, how effectively the hazard is controlled and supervised and the exposure to the hazard itself. The HSE rate the likelihood of injury as:

- High.
- Medium.
- Low.

Furthermore, the HSE confirm that risk can be defined as the combination of the severity of the hazard and the likelihood of occurrence and this is expressed as:

$$\text{Risk} = \text{Hazard severity} \times \text{the likelihood of occurrence}$$

but be warned that numbers are only numbers and do not mean that the event will not happen. It is just as probable that it will!

Assessing risk will also require the safety auditor to be aware of the accident situations regarding each activity as well as having an understanding of the task and role analysis information together with details of the agreed performance standards concerning those activities. When all risks have been identified an assessment can be made of the control measures used and their effectiveness.

ACCIDENT ANALYSIS

Having examined accident primary and secondary data it is a requirement of the audit to test how this is analysed and used within the decision-making process. An essential part of the analysis process is the auditing of the information provided by those involved in the accident or dangerous occurrence and those who were witness to the event. It is worth commenting upon the issues involved when gathering and analysing information gathered from people.

There are two common methods of collecting primary accident data for analysis; the interview and the questionnaire. There are other methods available such as experimentation, observation, self-recording and so on, but these are relatively sophisticated and rarely applicable to operational project conditions.

Choosing the appropriate method

The fundamental decision whether to use the questionnaire or interview is almost always concerned with resources and the scope of the problem:

- How much time and effort and manpower can be allocated to the project?
- Where are the sources of data located?
- What kind of data is needed?

The answers to these questions often mean that one of the methods is rejected and thus eases the decision to be made. Other choices can be eliminated by discussing appropriate advantages v. disadvantages whilst in more marginal situations, a choice is made by comparing direct comparison of the two methods.

Using the questionnaire

The advantages of this method are:

- It can be administered to groups of people at the same time thus saving time and expense.
- It can be used where the respondents are spread over a wide geographical area.
- The required answers on a questionnaire can be pre-structured thus making the task of analysing much easier.
- Questionnaires can be more reliable than interviews.
- Questionnaires can be made anonymous thus allowing for more freedom of response.
- Questionnaires give the respondent the opportunity to verify factual data.
- Questionnaires avoid interviewer 'error'.

The disadvantages of this method are:

- Questionnaires often suffer from a low response rate, which give rise to problems of representativeness. Respondents can also avoid answering specific questions.
- Difficult questions cannot be clarified by the respondent.
- Questionnaires are often seen to be impersonal and respondents do not feel committed to filling them in.
- There is no opportunity to ask extra questions on a questionnaire and therefore new data is missed. 'You only get answers to the questions asked.'

Conducting a formal interview

It is important to deal with some general principles of the interview before specific safety examples are discussed. The interview is a face to face verbal exchange, in which one person, the interviewer, attempts to elicit information or expressions of opinion, attitudes or belief from another person or persons.

The role of the interview is to act as a method for collecting details of the accident or dangerous occurrence. This may be used:

1. During the early stages of an investigation to help identify the problem areas and relevant dimensions, to suggest hypotheses and to reveal the natural frames of reference existing in the minds of the respondents.
2. When questionnaires are used, the interview may be employed to pre-test the questionnaire form.
3. As the main instrument of data collection.
4. To clarify findings which have emerged from other sources of data.

The advantages of these are:

- Interviews increase respondents commitment.
- They tend to be more valid, encouraging true to life answers.
- Interviews often surface data which would not come out in a questionnaire.
- They allow for clarification of difficult questions.

The disadvantages of these are:

- The problems of interviewer 'error' such as appearance, manner and style of asking questions.
- The problems of anonymity or the lack of it exist.
- Unstructured interviews often provide data in a form which is difficult to analyse.
- Interviews are time consuming.

The structured v. the unstructured type of interview

A structured interview is one in which the questions have been decided upon in advance of the interview. The questions are asked with the same wording and in the same order for each respondent. The essential feature of a structured interview is that the interviewer does not have the freedom to re-word questions, to introduce questions which seem especially applicable to the individual case, or to change the order of topics to conform to the interviewer's spontaneous sequence of ideas. In the unstructured interview, the interviewer technique is completely flexible and can vary from one respondent to the next.

Structured interviews:

- incorporate a basic principle of measurement (that of making information comparable from case to case);
- are more reliable;
- minimise errors due to question wording.

Unstructured interviews:

- permit standardisation of meanings, rather than the more superficial aspects of the stimulus situation (the question);
- are likely to be more valid in that they encourage more true-to-life responses;
- are more flexible.

Questions

For both interviews and questionnaires, it is necessary to relate questions to the problems under study. However, the discipline of refining questions and making sure that they are relevant can be difficult. For example, is a question pertaining to an employee's hobbies relevant to the study?

There are a few basic rules that can be applied to the framing of questions and these are:

1. Avoid ambiguous words or phrases.
2. Avoid long questions.
3. The questions should state as precisely as possible the time, place and context you want the respondent to assume.
4. Either make explicit all the alternatives the respondent could answer or none of them;
5. Avoid multiple questions.
6. Avoid leading questions.
7. Avoid rhetorical questions.
8. Avoid implied values.

It has been widely assumed that the interview is superior in many ways and must be used whenever resources permit. Certainly the interview should be used at the exploratory stages particularly in the area of accident investigation. Because of the seriousness of the accident investigation process and the legal requirements to make all workplaces safer such an interview may be referred to as a statement interview or formal interview. This is discussed below.

Statements or formal interview

If statements are considered as part of the operational plan it is important that they are taken as quickly as possible after the event and in normal circumstances should be taken within 24 hours of the accident happening. As early as 1932, psychologists have been able to show that a person is likely to forget half of what they originally perceived after 24 hours. To take a statement after this period of time can result in memory loss which may affect important detail, or other factors may influence the casualty thus distorting the facts. Further research has shown that a high status questioner can also influence a witness.

It should be understood that the taking of statements has a prime objective, and that is to ascertain facts relating to an accident or dangerous occurrence, which will facilitate the introduction of remedial measures which are specifically designed to prevent or reduce the incidence of those events happening in the future. The taking of statements should not be seen solely as a means of apportioning blame or deciding upon disciplinary action. The safety practitioner conducting such an interview may wish to obtain these facts in total confidence. Organisations should provide confidential phone lines for the reporting of dangerous occurrences. This practice should be recommended and form a part of the safety audit.

Disciplinary and management action must be examined as part of the safety audit as must the way in which information has been gathered from people.

EQUIPMENT AND SYSTEM INSPECTIONS

Pressure systems and vessels

On 1st July 1990, the Pressure Systems and Transportable Gas Containers Regulations, 1989 came into effect. These regulations consider safety of pressure systems not only at work but also in respect of:

- design;
- manufacture;
- import;
- supply.

Pressure systems now require the provision of design information and appropriate markings. All such systems will now require correct, safe installation. There are duties placed on employers regarding the competence and qualifications of persons who install and maintain such equipment.

A competent person is described as a person who must have certain attributes according to the complexity of the system and would be an engineer of Chartered or Incorporated status.

The safety auditor should:

- know what pressure systems are installed at each of your premises;
- know what the safe operating limits are;
- know where the inspection certificates are (either you should have them or your insurers);
- know who your 'competent' person is;
- have a schedule of your pressure system drawn up. (Although this is not a legal necessity it is good management practice.)

A system includes all associated pipe works, pressure parts and protective devices.

The regulations apply to you if:

- The company contains compressed gas (such as compressed air or liquefied gas) at a pressure greater than 0.5 bar (approximately 7 psi) above atmospheric pressure.

- There is steam where a pressure vessel exists and the plant is used by employees and/or self-employed persons.

Competent persons

Until the introduction of the pressure systems regulations, the term 'competent' had not been defined, but it is now accepted that a competent person is one who has such practical and theoretical knowledge and actual experience of the type of machinery. Also, he must be familiar with the plant which he has to examine in order to enable the detection of defects or weaknesses which it is the purpose of the examination to discover. Also, a competent person must be able to assess the importance in relation to the strength and functions of the particular machinery or plant he is examining.

In law, a person may also mean a company as well as an individual therefore it is quite acceptable for a competent person to mean a competent company. The following bodies may provide competent person services:

- A self-employed person.
- A partnership of individuals.
- A user company with its own in-house inspection department.
- An inspection organisation providing such a service to clients.

It was important to clarify the meaning of a competent person because the HASAWA embodies the basic principle that an employer is generally responsible for the work activities of employees. When an owner or user procures another company to provide competent person services, the contract is not made with individual examiners but with their employer, who is ultimately responsible for their work. Statutory inspection reports are normally signed by the individual inspector on behalf of their employer. The duties of a competent person are to:

- advise the owner or user of a pressure system which is covered by the regulations of a written scheme of inspection;
- certify or compile written schemes of inspection;
- carry out inspections and assess future usage of the system.

Organisations which specialise in providing advice on suitable competent persons are given at the end of this book under 'list of organisations'.

Lifting machinery and equipment

It is estimated that each year over 20,000 reportable injuries involve transport at work sites. Of these, some one-third involve lift trucks. Not only do lift trucks cause injury and death, they also damage equipment, goods being handled, buildings and fittings. From accident investigation it is found that lift-truck accidents usually involve a driver or operator who has not received correct training. The safety auditor must, therefore, confirm that all lift-truck drivers have received the following training:

- A basic course which provides the basic skills and knowledge necessary for safe operation, knowledge of the workplace and experience of any special needs and handling attachments.
- A specific course designed to give on-the-job experience under strict supervision.

Safety auditors should also satisfy themselves that the training provided conforms to the Approved Code of Practice for the Training of Operators of Rider-Operated Lift Trucks. They should also ensure that instructors who are to provide this training are themselves qualified and that they only give training on lift trucks on which they have been examined and certified. Instructors should also have a knowledge and understanding of the environment in which their trainees are to operate. All trainees should be tested in all skills and knowledge which are required for safe operation. An employer should keep a record of each employee who has received training in accordance with the Code.

Electrical equipment

On the 1st April 1990, the Electricity at Work Regulations, 1989 came into effect. These regulations provide for all aspects of electrical safety whilst at work. The regulations describe:

- An electrical system as one where all electrical equipment shares a common source of electrical energy.
- Electrical equipment as anything used or intended to be used to generate, control, distribute, use and anything else done with a supply of electricity.
- A circuit conductor as anything capable of conducting electrical energy.
- Danger as the risk of injury.
- Injury as death or personal injury from electric shock, burn, fire, explosion or arcing initiated by and associated with electrical energy.

All electrical systems shall be constructed and regularly maintained so as to prevent injury or give rise to danger. In addition, all work activities at or near electrical installations shall be carried in a safe manner and safety equipment provided shall be suitable for the purpose intended.

It is important that all electrical sockets, cables and electrical equipment are regularly checked and maintained by a qualified electrician. Care should be exercised when extension cables are used and advice sought where appropriate. It is the duty and responsibility of a safety manager to ensure that all technical aspects of health and safety are carried out by properly trained and qualified staff. The safety auditor must carry out a visual inspection of all electrical equipment maintenance records and fittings and satisfy himself that all installations, servicing and maintenance has been conducted by competent and appropriately qualified staff. A note must be made of any delays in servicing and maintenance procedures and reasons established.

Construction and building issues

Construction and building sites are recognised as the most hazardous of all. This statement is based upon published accident information. According to these statistics, the construction industry employs under 10 per cent of the working population yet provides 15 per cent of reported accidents and over 30 per cent of the fatalities. Site management is a specialist task and personnel managers with a responsibility for health and safety matters in the construction industry must seek appropriate advice and ensure that:

- The site manager is qualified to manage the health and safety requirements.
- That appropriately qualified and trained staff are appointed.
- That scaffolding, lifting gear and other apparatus conform to current codes of practice.
- That resources are available for the provision of safety clothing and equipment where appropriate and that they are used correctly.
- That a building site education and training programme is available.
- That safety publicity is given a high priority.

There are a number of issues to consider. For example, we know that working at heights is a hazardous part of the construction industry and it is usually human error rather than equipment failure which is the major contributory factor in this type of accident. It is important therefore to ensure when workers are 'above ground' that:

- the right ladder is used;
- the ladder is set up correctly;
- appropriate ancillary equipment is set up correctly;
- equipment is inspected before use;
- ascents and descents are carried out correctly;
- weight capacities of the equipment are known and understood;
- all equipment is checked at the start of each working day;
- lifting gear engines are warmed up before use;

- clearance from electrical cables exists and is maintained;
- safety equipment is not worn/used;
- high wind loading procedures are in place.

Here it is important to have direct access to the task analysis and performance standards information. Each task on the building or construction site can then be assessed against this criteria and an assessment made of the safety management processes involved with subcontractors, visitors to the site and the general public.

Maintenance procedures

All machinery, safety clothing and equipment must be regularly maintained and inspected if they are to remain effective. Machinery safety has been highlighted more than any other by enforcement, case law or on civil liability. It is now accepted that the hazard-free machine cannot be produced, therefore it is a requirement to guard or fence machinery in such a way as to minimise the risk of injury to the worker or operator. Most case law, therefore, has tended to concentrate upon the guarding and fencing of machinery rather than on the designers, manufacturers and suppliers of the equipment.

The following boundaries have been established by case law in relation to the guarding of machinery:

- Only machinery which is in use has to be guarded or fenced.
- The purpose of guarding/fencing is to prevent employees and/or operators coming into contact with the machine.
- Transmission machinery and prime movers are deemed to be dangerous so statutory duties relating to them are considered absolute whereas other parts of the machinery are presumed to be safe unless shown to be hazardous.
- Only machines used in the factory process are deemed to be machinery.
- Mobile as well as static machinery is covered by statute.
- Machinery, although guarded, may require fencing when working on material or where it is in close proximity to other machinery.
- Hand-held tools are not regarded as machinery.
- There is no requirement to fence where an unforeseen danger is created by the interaction of a moving part of a machine and a stationary object.
- If a machine is dangerous it must be fenced, putting up warning signs alone is not sufficient.

The Health and Safety Executive have classified certain parts of machinery which they regard as being dangerous and because of this should be securely fenced. The parts are:

- revolving shafts, mandrels, bars and spindles;
- in-running teeth between pairs of rotating parts such as gears;

- in-running nips of the belt and pulley sort such as those used in conveyor belts;
- projections on rotating parts;
- discontinuous rotating parts such as fan blades;
- revolving beaters, spiked cylinders and revolving drums;
- revolving mixer arms in casing such as dough mixers;
- revolving worms and spirals in casings such as mincers or extruders;
- revolving high speed cages in casings such as centrifuges;
- abrasive wheels;
- revolving cutting tools;
- reciprocating tools and dies such as power presses and drop stamps;
- reciprocating knives, blades and saws;
- closing nips between platen motions such as printing machines;
- projecting belt fasteners and fast running belts;
- nips between connecting rods, rotating wheels, cranks and/or discs;
- traps arising from traversing carriages such as metal planing machines.

Whilst there is a vast amount of case law concerning dangerous machinery and lists provided by the Health and Safety Executive, it is important for safety auditors to:

- identify dangerous machinery;
- ensure that dangerous machinery is guarded and/or fenced;
- ensure that employees are aware of the dangers;
- ensure that appropriate education, training and publicity is given to employees about these dangers;
- ensure that all guards/fences are regularly inspected and maintained in good working order.

HSE CODES OF PRACTICE

Safety audit processes should examine each activity undertaken in the workplace and a note should be made of each activity which has an HSE Code of Practice issued or where guidelines or advice are issued either by the HSE, the manufacturer and or the relevant trade association. Furthermore it is necessary to check whether management are aware of these and that they are considered within the decision-making process.

PROFESSIONAL CODES OF PRACTICE

In addition to those points discussed in the previous paragraph there are also professional codes of practice which are usually a part of an employee's

	Satisfactory	Not satisfactory
Accident costing and evaluation Costing of accidents Valuation methodology Economic rates of return Accident investigation methodology Data collection processes Witness statements Interviewing techniques Disciplinary advice Accident records		
Medical matters Medical examination data First-aid treatment registers Numbers of first-aid personnel RIDDOR information RIDDOR management Qualifications of first aiders Siting of first-aid boxes Access to first-aid boxes First-aid equipment and box contents Rest room facilities		
Welfare Sanitary facilities (quantity) Sanitary facilities (condition) Sanitary facilities (soaps, papers and towels) Kitchen facilities (quantity) Kitchen facilities (condition) Kitchen facilities (provision of cleaning materials) Provision of hot and cold running water Provision of catering facilities Condition of mess/eating halls Cleanliness and hygiene standards Food hygiene and storage Stress management facilities Alcohol and drug advisory services		
Hazard and risk assessments Hazard assessment schedules Risk assessment schedules Hazard assessment methodology Risk assessment methodology		
Accident analysis Accident analysis practices Accident analysis methodology Use of information generated		
Codes of practice Use of HSE Codes of Practice Use of manufacturers' guideline notes Use of trade association notes and guidelines Professional codes of practice		

Figure 7.5 A safety practices safety audit summary checklist

professional qualifications. For example, a member of the Institution of Civil Engineers will be additionally bound by the codes of practice of this particular professional body and members will be subject to the disciplinary procedures so described. The safety auditor should not ignore these additional forms of protection when conducting the safety audit.

An example of a safety practices summary sheet is given in Fig. 7.5.

8 AUDITING SAFETY PROGRAMMES

Here, the safety audit will examine past remedial strategies for efficiency and effectiveness and involving the four basic elements of the safety mix:

- **Safety auditing enforcement strategies**
- **Safety auditing engineering remedial measures**
- **An environmental safety audit**
- **Education, training and publicity**

The safety mix forms a main feature of the accident reduction and prevention process and successful safety programmes are based upon a mix of some or all of the four main features, i.e. enforcement, engineering, environmental improvements and education (including training and publicity).

SAFETY AUDITING ENFORCEMENT STRATEGIES

Care should always be taken to ensure at safety audit that the strategy does not rely solely upon the enforcement aspect of the mix as the only means of providing effective accident prevention. This will not be effective and further information is given in Chapter 3 of the Handbook of Safety Management.

Whilst the manager might consider the planning, implementation, monitoring and evaluation of the safety mix programme within specified areas of operation, enforcement would consider those sets of rules and regulations which are enforceable by law as a feature of the safety mix. Such measures would include:

- common law relating to health and safety;
- statutory laws;
- regulations and other statutory instruments provided for under the law;
- local bye-laws and other local government regulations;
- HSE codes of practice and guidelines;
- manufacturers' advice notes;
- guidance notes issued by trade associations;
- professional codes of practice;
- British and international safety standards;
- general disciplinary codes of practice and rules.

The safety auditor would need to identify at least one significant accident reduction strategy implemented since the previous audit as well as the enforcement features within the mix and those which are relevant, and note how and when they were implemented. These would need to be compared with those actually used. Cross-reference should be made against the ten features listed above and a note made of those which could have been used and were not. Questions would then need to be asked of management as to the reasons for this in order to put the information into perspective with the rest of the safety audit data.

SAFETY AUDITING ENGINEERING REMEDIAL MEASURES

An important feature of the safety mix is the contribution that engineering makes to the organisational accident reduction and prevention strategies. Such measures would include:

- improved machinery design and manufacture;
- advancements in technology;
- automation requiring less human dependency;
- improved equipment design and production;
- better safety equipment and manufacture;
- improved servicing cycles;
- better maintenance programmes;
- improved engineering skills and training;
- improved construction techniques and materials;
- better engineering testing and operating procedures.

These would lead naturally to safer working practices. Here the safety auditor would need to identify all equipment sited in or around the area subject to audit and a note would have to be made of the condition of each piece of equipment and machinery. Information would be required concerning operating procedures and observations would need to be conducted to satisfy the safety auditor that such equipment was being used in accordance with the manufacturers' guidelines. A note would be made of all ancillary equipment worn or used which was specifically designed and issued for accident reduction and prevention purposes.

At this stage of the audit it would be necessary to obtain all servicing and maintenance records to ensure that equipment and machinery is regularly maintained in accordance with manufacturers' instructions. Any changes to machinery or equipment including servicing or maintenance schedules should be noted and reasons obtained for this from senior management.

AN ENVIRONMENTAL SAFETY AUDIT

This feature of the safety mix considers the working environment in which the employee works and moves and also includes those welfare issues discussed in Chapter 7. One should make a note of the general environment before moving to the specific working environment of the employee(s) being audited. The general environment would include such features as:

- temperature;
- noise;
- vibration;
- air pollution;
- cleanliness;
- hygiene facilities;
- fire procedures and facilities;
- first-aid facilities;
- radiation;
- manual handling facilities;
- general welfare;
- alcohol and drugs;
- stress management considerations;
- control of hazardous and dangerous materials and substances;
- equipment and machinery locations.

Having examined the general environmental conditions it is important to move to the more specific working environments of employees operating in higher risk areas moving progressively to the low risk category of worker. Extensive use will need to be made of internal records, approved codes of practice and guidelines, manufacturers' advice notes and other information issued by trades associations and professional institutions.

EDUCATION, TRAINING AND PUBLICITY

Safety education is a most useful feature of the safety mix and one which has great operation accident reduction and prevention benefits. It is worth discussing some of the more important issues here. Short-term based safety education may be required at various stages to educate the general public or workforce about a potential or actual hazard, or to provide specific safety training programmes. It is not possible here to discuss the psychology of learning and managers and safety auditors should refer to Chapter 10 of the Handbook of Safety Management for this information.

The safety auditor will need to examine schemes and plans announced by the organisation since the previous safety audit which attempt to educate its workforce in such a way as to develop skills, knowledge and safety behaviour.

Auditing safety training

Safety training is a very broad term which includes any activity that is to improve an individual's performance, increase his contribution to organisational effectiveness and to reduce or prevent accidents from happening. There are many approaches to safety training within organisations varying from the systematic approach to vague ideas about things people ought to know about in order to do their job safely. Safety management development is an area often singled out from the general area of training in order to identify the process of improving safety practitioner's knowledge, understanding and abilities with the aim of improving the competence of the safety management team. Safety management development programmes are often more individually tailored than, for example, training programmes for shop floor workers. Whilst there may be differences in the type of learning objectives and in the time scale involved, it is felt that the same principles apply throughout the general area of safety training and safety management development. The management of the safety training function is a fourfold process. These would be subject to safety audit:

1. Assessing the need – determining the safety training requirements for all types of staff, deciding priorities and defining standards.
2. Programming – plans and procedures aimed at fulfilling needs in terms of the policy on internal or external courses; individual development plans; deciding upon the techniques most appropriate to each type of safety training.
3. Organising – how best to use the staff, finance and resources available for safety training purposes.
4. Evaluation – how well the results meet the original needs; budgetary control of resources.

A major problem concerning safety training is that often it is carried out with unclear objectives and a failure to diagnose real training needs which may lead to the whole value of safety training being questioned. In this chapter, the following shall be discussed:

- Objectives of safety training.
- Approaches to safety training.
- Safety training methods.
- Evaluation of safety training.

General aspects to consider when using safety training as remedial strategy are also discussed later in the chapter.

The objective of safety training can be stated quite simply as an activity designed to improve organisational effectiveness and safety. However, this overall general objective needs to be broken down into more specific goals. The area of safety training can be divided into two major activities:

- Safety training.
- Safety development.

This division is based on the objectives which are to be achieved. If the activity is specific, factual, short-term and includes only a narrow range of material (e.g. new safety regulations such as COSHH) then it can be defined as training. If the activity is concerned with developing abilities, skills and changing attitudes in the broad safety management sense then it is often termed safety development. The division between training and development is very unclear and many organisations carry out safety management training which could well be described as safety development and *vice versa*.

A typical safety training objective might be to increase a worker's knowledge and understanding of a particular subject such as the HASAWA, COSHH regulations or internal fire alarm procedures. Or, to develop skills both physical and mental, safe use of equipment, problem solving, etc. Attitudes may be changed regarding safety procedures, policies and practice, and rewards can be made to individuals for good, safe work. A safety training objective might be to maintain the labour force in good, safe condition. Although these objectives are extremely general it is important that if training is to be effective the objectives need to be thoroughly considered and related to the objectives of the organisation. Safety training for safety training's sake frequently occurs because safety objectives have not been drawn up in any meaningful way. Safety training requires objectives to be:

- specific;
- realistic;
- measurable.

There are several ways in which an organisation may approach the activity of safety training, varying from ways which develop training policies to those which operate on an *ad hoc* and unsystematic basis. In order more clearly to understand the question of approaches to safety training there are four activities of safety training activity which would be subject of the safety audit process. These are the:

1. Administrative approach;
2. Individual approach;
3. Organisational development approach;
4. Systematic approach.

These four categories are now discussed.

The administrative approach

Here, safety training is approached with an administrative bias as the name would suggest. Courses are organised and people are fed into them by their departments or by the safety manager. The disadvantage is that the diagnosis of their safety training needs is very weak and whole groups of jobs, such as all supervisors, are sent on courses whether or not they are appropriate. The evaluation of this approach tends to be weak as success is often seen as numbers trained and the amount of grants received or levies not paid. Whether or not the training has improved either the individual's safety performance or the organisation's safety effectiveness seems of secondary importance.

The individual approach

As the name suggests, safety training is given based upon the needs of the individual. It is normally instigated by an individual requesting a training course and for the organisation to support this request. The safety training requested may well meet the needs of the individual but these may not be linked to the organisational needs. Often the safety training is carried out on an external basis by consultants or safety organisations and the courses may attract people from different industries and backgrounds, with the safety training consequently lacking direct relevance. This technique tends to be weak in terms of evaluation as it is based on the subjective assessment of the individual who attended the safety course, seminar, conference, etc. This approach is prevalent in many organisations and it is difficult to justify when considering those safety training objectives outlined above.

The organisational development approach

This approach lays great emphasis on the diagnosis stage of the safety training process. The specialist and relevant managers spend considerable time attempting to define the real problem and what real safety training requirements should be and whether it is a safety training issue at all. The primary concern of this approach to safety training is to concentrate on improving health and safety processes that operate within the organisation. This approach is also concerned very much with organisational change and diagnosing needs in this area rather than concentrating on the traditional areas of safety training for a particular industry.

Internal and self-managed safety training is stressed to a great extent in this approach. Management should be involved in planning and carrying out their own safety training by using and learning from their own experiences. Much of the safety training under this category is concerned with groups of workers on vertical, diagonal or horizontal slices through the organisation. Safety training should be developed at a group level if major changes are to be made in the organisation particularly if safety performance is to be improved.

The systematic approach

This follows a rigorous approach to all aspects of the total safety training process. For convenience it can be divided into four important steps:

1. The diagnosis of safety training needs;
2. The objective setting phase;
3. The design of the safety training programme;
4. Evaluation of the safety training programme.

This approach to safety training is time consuming and because of this is often neglected by organisations.

Diagnosis of the safety training needs

To define specific training needs, it is necessary to begin at the macro level and then come downwards towards the individual safety training needs at the micro-level. In order to start at the macro level, it is important to know the organisation's objectives so that any shortcomings in safety performance may be identified. Areas such as accident rates or dangerous occurrences should be examined to see if there is any deterioration in performance. It should be remembered that detailed accident analysis is required if contributory factors are to be identified. These factors may well have little to do with training (e.g. the 0.25 enforcement sector within the safety mix may need to be increased), but assuming that a safety training need has been identified it is necessary to consider at what occupational level safety training is required and in particular, what individual safety training needs exist.

There are a number of questions that need to be answered at this stage:

- for which occupations is safety training required to cater for current weaknesses or for future accident reduction needs?
- how many people will need training?
- what are the critical areas?
- where can the best return on the investment be made?
- what resources and constraints will affect these questions?

Having answered these questions it is then necessary to examine the job or occupations selected as the priority area and decide whether the present system could be changed or reorganised to obviate the need for safety training. Assuming that it is accepted that a safety requirement exists, as is quite often the case, the individual job needs to be analysed in terms of their objectives, activities, component parts, skills, knowledge necessary to carry out the job safely and effectively. The next step is the final one before considering the objective setting sequence. As a detailed analysis of the skills and knowledge required to perform the task has been drawn up it is necessary next to detail the skills and knowledge already possessed by the target population. The difference between the skills and knowledge already possessed and those required to do the job effectively and safely is known as the safety training gap.

Objective setting

This part of the systematic safety training process is really attempting to answer the question: 'what do we want the trainee to be able to do when training is complete?' The safety auditor will also require information concerning performance standards expected and achieved.

This means that criteria should be developed which are both quantifiable and measurable. This can be a difficult process as often safety training programmes are attempting to achieve vague and indefinable changes such as an improvement in attitude or behaviour. However, wherever possible it is worthwhile to attempt to develop criteria which can be assessed as this clarifies the process of programme design as well as the evaluation process.

Objective setting tends to be easier where straight knowledge and physical skills are involved but become more difficult when concerned with safety management training which goes further than straight knowledge inputs.

Designing the safety training programme

This stage contains two steps:

1. To decide the content of the training activity;
2. To plan the method and sequence of the programme.

The content of the safety programme should stem from the training gap identified earlier and will obviously link closely with the objectives that are defined. The second step is probably more difficult as it involves making decisions about the sequence of learning, where it will take place, who will carry it out and what resources are available, etc. The answers to these questions will depend on many factors such as the level of motivation of the trainees, the degree of complexity of the material and the safety manager's (or his trainer's) beliefs concerning learning theories and so on. The planning of the safety training programme can only be carried out within the context of a particular situation and therefore the many factors that need to be considered cannot be detailed here as they need to be referred to specific cases in point. A further important point concerns the teaching methods to be employed. There are many methods available and these will be discussed later in this chapter.

Evaluation

This part of the process is usually ignored by many safety practitioners yet if properly organised can answer important questions such as:

- has the training achieved its objective?
- has it satisfied the training need?
- do the benefits justify the costs?

Safety training methods

There are several methods available for transferring safety information to others. At the basic level, safety training methods can be divided into on-the-job and off-the-job, and again the decision as to which is most appropriate will depend upon the situation in question. These two types of safety training are now discussed.

On-the-job safety training

The safety auditor will consider four of the methods of on-the-job training to consider:

1. Learning from another skilled person.
2. Taking part in project work.
3. Taking part in a job rotation programme.
4. Individual coaching.

Learning from another person is a very common method and is used for simple tasks where the risk factors are very low. The method involves a new worker sitting with an experienced operator and learning the job by observation and by trial and error. Although this can sometimes be an inexpensive method of learning provided that the operator is familiar with safety procedures and has a good safe work record. It will be necessary to quantify how many new workers have been trained this way since the last safety audit and to ascertain who provided this form of training. Job training performance standards and safety performance appraisal summary sheets will also be examined.

Taking part in a project is used where employees are required to have reached a certain academic and intellectual level such as graduate trainees who are allocated specific problems to work on, understand problems and develop solutions. This can be an effective safety training method as long as the projects are chosen with the trainees' needs in mind and encouragement and feedback are given on a systematic basis. Again the safety auditor must examine the projects undertaken and examine safety performance standard criteria.

Taking part in a job rotation programme involves moving a trainee around either between departments or between various specific jobs. Job rotation is a useful exercise where the operator is to be promoted to a supervisory position and a knowledge of safety procedures of the various jobs under their supervision is necessary. The idea is to familiarise trainees with the various functions within the organisation and to grasp the overall safety requirements of the organisation. These activities need to be quantified and audited in the same way as above.

Individual coaching is an extremely useful method of safety training which develops from the manager/subordinate relationship. The safety auditor must identify how often this method has been used and discover who carried out the coaching. It is important to examine objectives and safety performance standards and to look at the safety performance appraisal cycles used. It would be normal to increase safety performance appraisals for new trainees.

Off-the-job safety training

There are many methods of safety training which are available and the reader will have experience of many of them. The methods need not be discussed in great detail but are merely listed, as advantages and disadvantages of each type can be drawn from readers' experience. The more common safety training methods are:

- Lectures.
- Seminars.
- Closed circuit television.
- Interactive video.
- Programmed learning.
- Films.
- Cases.
- Role play.
- Business games.
- Experimental learning techniques.
- Conferences.

All of these methods, if used, would be examined by the safety auditor together with assessments of the competencies and skills of all lecturers and course content and method of delivery.

Safety publicity

Safety publicity forms an important part of the education process within the safety mix. It is not necessarily confined specifically to campaigns and can include public relations and low key educational programmes designed specifically to inform rather than to persuade or change behaviour. Safety publicity can be an ideal way of promoting the company safety image whilst at the same time being an effective tool against accidents. It can be expensive and does require careful planning if it is to be used. Researchers have found, however, that, used in isolation, safety publicity is not very cost effective and should therefore be considered as part of an overall safety related programme. At the early planning stages it is important to identify the key areas of the safety mix which are to be employed, and if safety publicity is to form part of this it is essential that its role is clearly defined. In the past many organisations have been over enthusiastic about publicity strategies because they advertise what is being done when in fact little or nothing may be being achieved at all. These exercises tend to be public relations exercises rather than carefully planned campaigns specifically designed to contribute to the accident reduction efforts of an organisation.

From the analysis of accident data areas for remedial action will be identified. Assuming that an effective strategy will be the use of a safety publicity campaign and sufficient financial resources are available then the following will need to be considered by the safety auditor:

- Has a similar exercise been conducted either internally or externally since the last safety audit?
- What form did the campaign take? What creative ideas were discussed?
- Who were the target audience? What were the campaign objectives?
- Obtain the communications brief.
- Were external or other agencies involved?
- What, if any, were the non-media counter-measures?
- Focus on the implementation strategy for safety audit purposes.
- What refinements were made?
- How was the campaign or programme implemented?
- What monitoring occurred?
- By whom and when was the campaign review conducted?

Where safety literature or posters are used, copies of these should be obtained and a note made of how leaflets were distributed and where posters were located. The safety auditor should take a random sample of employees and ascertain their skills, knowledge and/or behaviour which was the subject of the leaflets or posters.
A summary checklist of the safety mix audit is given in Fig. 8.1.

	Satisfactory	Not satisfactory
Enforcement		
Acts of Parliament		
(list those relevant to the industry)		
Case law		
Regulations		
Statutory instruments		
Orders		
HSE codes of practice		
Manufacturers' advice notes		
Trade association guidelines		
Company guidelines and policy		
Profession codes of practice and conduct		
Standard disciplinary procedures and practice		
Work and task schedules		
Work methods and procedures		
Engineering		
Safe design of machinery		
Safe design of equipment		
Safety clothing		
Safety equipment		
No authorised modifications to:		
Machinery		
Equipment		
Materials		
Machinery maintenance schedules		
Machinery repair schedules		
Equipment serviceability and maintenance		
Safety equipment storing, issue and repair		
Safety equipment maintenance schedules		
Safe operating procedures		
Safety limitations of equipment and machinery schedules		
Environmental matters		
General workshop/work area cleanliness		
General building condition		
Heating		
Lighting		
Ventilation		
Dust and fumes		
Welfare		
Clean and tidy work areas		
Sanitary conditions		
Hygiene		
Stress management		
Alcohol and drug (no smoking areas)		
Noise and vibration		
COSHH		

Figure 8.1 A summary of the safety mix audit checklist

	Satisfactory	Not satisfactory
Education Company safety education policy Professional qualifications Competence levels Provision of safety journals Provision of a safety library Other safety literature Membership of professional societies and institutions		
Training Induction training to safety Continuation safety training		
Safety training facilities:		
Classrooms Equipment Lecturing staff Learning materials Provision of support from outside agencies		
Curriculum Delivery Instructor competences		
Advanced safety training		
Attendance at IOSH/IIRSM/other professional training Attendance at trade association courses and seminars		
Company encouragement for safety training		
Publicity Internal company staff relations Internal company communications External public relations Safety related exhibitions Safety promotional policies and programmes Safety literature production Internally produced leaflets and safety publications Internally produced safety posters External materials Use of language and appropriateness to target group Company involvement in national campaigns HSE publications issue Use of professional/trade association materials		
Use of company accident data and local issues Involvement in all levels of company staff		
Use of professional agencies		

Figure 8.1 cont. A summary of the safety mix audit checklist

9 THE SAFETY AUDIT DOCUMENT

In this chapter we will look at the following:

- **Introduction**
- **Quantifiable data**
- **Using professional judgement**
- **Designing the safety audit checklist**
- **Measuring some aspects of performance**
- **Some safety performance calculations**

INTRODUCTION

In Figs. 5.14, 6.2, 7.5 and 8.1 the safety audit data gathering forms require only a satisfactory or not satisfactory record to be made. This is because they are merely summary sheets of the main activities to be examined. On the actual safety audit document it will be necessary to identify:

- All data which is quantifiable.
- All observed criteria, checked against that which is quantifiable.
- All observed data which is subjective in nature, checked against professional judgement.
- Comments on whether or not the results are satisfactory.

This information will allow the safety audit report to be written and will provide information against each activity and task. It is usual for many commercially available safety audit documents to award points or scores for the various categories within the safety audit. This should be avoided. It has been said earlier how this can give management a false sense of security and it is worth re-emphasising here. No scores are used in the safety audit document apart from a 0 or 1 which could be required for the computerisation of the data at a later stage and should not be interpreted as a form of scoring.

QUANTIFIABLE DATA

Obtaining this is the most important part of the safety auditing process. It is usually also an opportunity to examine the safety management process and its ability to quantify objectives, tasks, roles and performance standards not only of itself but for the remainder of the workforce. Where something is said to be quantifiable it is usually beyond question. For example, a performance standard might be to reduce accidents in the 'print room' by 15 per cent. This can be checked (assuming that accurate records are kept) by calculating the difference between, say, the number of accidents since the last safety audit and this one. If a 15 per cent reduction has been achieved then this would be regarded as meeting the objective and would be classified as being satisfactory. If, for example, the accidents had been reduced only by 14 per cent then this is not meeting the objective and questions would need to be asked as to why this was so. In smaller organisations, the objective might be to reduce accidents by three (rather than using percentages). This, too, can be checked at safety audit in exactly the same way. If objectives or performance standards are not being met, then this could be due to:

- inappropriate or no appraisal interviews;
- no, inappropriate or unachievable performance standards;
- inappropriate or no feedback.

The safety auditor therefore has to identify all quantifiable performance standards, tasks, roles, and objectives at the commencement of the safety audit. At the same time, it is necessary to obtain all appropriate available primary and secondary data regarding each of the functions. This will allow the necessary calculations to be made which will test whether these standards have been achieved.

USING PROFESSIONAL JUDGEMENT

There will be some areas which are extremely difficult to quantify. For example, what might appear to be a source of danger to one person might not be to another. This brings us to the subjective areas of performance standards and in order to resolve these differences of opinion some form of judgement is made. To help in this, accident data is the most useful. An opinion as to the danger of some hazard or risk could be examined against what is already known. Can this fear of danger be justified? By examining closely the company accident and dangerous occurrence data it should be established whether this is the case. If no accidents or dangerous occurrences have been recorded then the hazard or risk can be down-graded in terms of priorities for action and a note made accordingly. If evidence is found then the hazard or risk can be confirmed and appropriate steps introduced to deal with the matter.

Bloggs Manufacturing Ltd Sheet 1 of 9 Safety Audit Checklist Carried out by: Date safety audit commenced: Date safety audit finished: Division/Department/Unit/Section/Individual: Safety audit of policy/procedures/practice/programmes/safety mix Location of safety audit: Signature of safety auditor: .... Date: .... **Safety tasks**	Quantifiable	Observed against quantifiable	Professional judgement	Satisfactory	Unsatisfactory
Preparatory information: Previous safety audit information Accident records Accident costing information Company statement on health and safety policy Accident frequency rates Safety audit aims and objectives Safety audit performance standards Discussions held with senior management Discussions held with supervisors Staff appreciation of the aims and objectives of the safety audit Access to all data and records Identification of all data access points The order of the safety audit **Legal requirements:** List all Acts of Parliament which are applicable to the industry to which the company belongs List all Regulations, Orders and other Statutory Instruments which are applicable to the industry to which the company belongs Check health and safety policy document for the following: * when it was last updated * how it is communicated to staff * language used * presentation and general style * access to the document by all staff * reminders given to staff since the last safety audit * knowledge of all staff as to the content of the policy statement Examine the company insurance documentation Examine fire equipment Examine fire procedures Examine fire signing and access points Examine fire equipment maintenance records Check visit records of Fire Officer Check first-aid facilities Check qualifications of first aiders Check siting and contents of first-aid boxes Examine first-aid (treatment) records Examine first-aid room/recovery room Obtain medical record summary sheets of injuries and related illnesses List numbers of medical/first-aid staff available					

Figure 9.1 A section from a safety audit checklist

Bloggs Manufacturing Ltd Sheet 2 of 9 Safety Audit Checklist Carried out by: Date safety audit commenced: Date safety audit finished: Division/Department/Unit/Section/Individual: Safety audit of policy/procedures/practice/programmes/safety mix Location of safety audit: Signature of safety auditor: .... Date: .... **Safety tasks**	Quantifiable	Observed against quantifiable	Professional judgement	Satisfactory	Unsatisfactory
Obtain minutes of all safety committee meetings held since the last safety audit List the number of safety representatives These are just a sample of the legal requirements which are necessary. You should now list all other legal requirements which are relevant to your particular industry so that they can be systematically checked. **Statement of objectives** Obtain statements of objectives from: * the organisation * regions (where appropriate) * divisions (if appropriate) * departments * sections * units/teams * individuals Obtain safety performance standards from: * the organisation * regions (where applicable) * divisions (if appropriate) * departments * units/teams * individuals Examine whether objectives have been met by: * organisations * regions (where applicable) * divisions (if appropriate) * departments * units/teams * individuals Examine whether safety performance standards have been met by: * organisations * regions (where appropriate) * divisions (if applicable) * departments * units/teams * individuals Identify shortfalls in objective/performance standards separately on the safety audit comment sheet					

Figure 9.1 cont. A section from a safety audit checklist

Bloggs Manufacturing Ltd Sheet 3 of 9 Safety Audit Checklist Carried out by: Date safety audit commenced: Date safety audit finished: Division/Department/Unit/Section/Individual: Safety audit of policy/procedures/practice/programmes/safety mix Location of safety audit: Signature of safety auditor: .... Date: .... **Safety tasks**	Quantifiable	Observed against quantifiable	Professional judgement	Satisfactory	Unsatisfactory
Budgetary provision: Examine the safety budget estimates and compare with those of the previous safety audit Examine whether safety budget estimates reflect aims and objectives of the: * organisation * regions (where appropriate) * divisions (if appropriate) * departments * units/teams * individuals Examine whether the safety budget complements the safety performance standards of the: * organisation * regions (where appropriate) * divisions (if appropriate) * departments * units/teams * individuals Examine the safety budget expenditure records for priorities and action Examine safety budget under/overspend Examine budgetary control mechanisms **Staff:** Check unit/team/individual role functions Examine unit/team/individual safety performance standards Examine performance review functions Examine performance review assessments/changes Examine new staff separately Check new staff induction course mechanisms Obtain/unit/team/individual task analysis details Check senior management role in setting and evaluation of safety performance standards Check the involvement of supervisors in safety performance standard setting and assessment Check the involvement of trades unions in the setting of safety performance standards/safety objective setting Examine the role of safety committees and safety representatives in the safety performance mechanisms (where appropriate)					

Figure 9.1 cont. A section from a safety audit checklist

Bloggs Manufacturing Ltd Sheet 4 of 9 Safety Audit Checklist Carried out by: Date safety audit commenced: Date safety audit finished: Division/Department/Unit/Section/Individual: Safety audit of policy/procedures/practice/programmes/safety mix Location of safety audit: Signature of safety auditor: .... Date: .... **Safety tasks**	Quantifiable	Observed against quantifiable	Professional judgement	Satisfactory	Unsatisfactory
Organisational structure: Examine hierarchical structures relating to: * regions (where appropriate) * divisions (if appropriate) * departments * units/teams * individuals Identify key features which support/do not support the safety performance standards/safety objectives of: * regions (where appropriate) * divisions (if appropriate) * departments * units/teams * individuals **Control of hazardous and dangerous substances and materials:** List all hazardous and dangerous materials which are brought into the workplace Check the safety information on the labels Identify compliance with this information and other information and data provided by the HSE/Trade Associations etc. Check Part 1A1 of the approved list issued under the Classification Packaging and Labelling of Dangerous Substances Regulations 1984; Identify anything listed as toxic, corrosive, harmful or irritant. Identify where and in what circumstances these substances are used, handled, generated and released Identify whether the substance form changes during or after use Examine current measures to protect and control exposure Identify who is affected by any exposure Identify the likelihood of substances being inhaled Is there a likelihood of skin contamination and what measures are in place to protect those who are likely to be affected Examine safety clothing for: * appropriateness * condition * wearing rate Examine delivery and storage facilities Examine issuing measures and facilities Question users for their comments (quantifiable questions wherever possible)					

Figure 9.1 cont. A section from a safety audit checklist

Bloggs Manufacturing Ltd Sheet 5 of 9 Safety Audit Checklist Carried out by: Date safety audit commenced: Date safety audit finished: Division/Department/Unit/Section/Individual: Safety audit of policy/procedures/practice/programmes/safety mix Location of safety audit: Signature of safety auditor: .... Date:	Quantifiable	Observed against quantifiable	Professional judgement	Satisfactory	Unsatisfactory
Safety tasks					
First aid: Examine first-aid records from: * regions (where appropriate) * divisions (if appropriate) * departments * units/teams * individuals Examine facilities at these locations for compliance with Section 2(1) of the HASAWA 1974 Examine first-aid equipment at all locations Test that provision is adequate (see figure 5.3) Test first-aid training facilities and availability **Heating:** Examine the temperatures at various times of the day i.e. * first thing in the morning * mid-morning * mid-afternoon Check: * Offices * Factory work areas * Steam rooms (where applicable) * Other rooms where artificial humidity is generated) Temperatures should conform to those given in figure 5.4 **Lighting and electrical equipment:** Refer to the FA 1961, the OS & RP A 1963 and ensure that: * lights and electrical apparatus are regularly maintained * lights and electrical appliances are in good working order * lighting conforms to the requirements listed in figure 5.5 * electrical equipment is maintained and used according to the manufacturers instructions Examine lighting conditions throughout the working areas Examine external lighting conditions at night (or darkness) for: * adequacy * siting (drivers should not be blinded for example) * lighting type * maintenance Examine competencies of maintenance engineers and that any repairs conform to approved standards					

Figure 9.1 cont. A section from a safety audit checklist

Bloggs Manufacturing Ltd Sheet 6 of 9 Safety Audit Checklist Carried out by: Date safety audit commenced: Date safety audit finished: Division/Department/Unit/Section/Individual: Safety audit of policy/procedures/practice/programmes/safety mix Location of safety audit: Signature of safety auditor: .... Date: .... <div align="center">**Safety tasks**</div>	Quantifiable	Observed against quantifiable	Professional judgement	Satisfactory	Unsatisfactory
Ventilation: Examine air purification and/or ventilation facilities in: * offices * all working areas * steam rooms * other places of work where fumes, dust and other impurities are evident Check the legal requirements and conformity to these (see figure 5.6) Check the provision of safety clothing and equipment Check that these are being used in accordance with company policy Examine current company policy for adequacy Examine maintenance records of safety equipment/clothing Identify any weaknesses in maintenance schedules Test the atmosphere in all working locations and measure for pollution **Noise and vibration:** Examine compliance with Noise at Work Regulations, 1989 Examine company policies,procedures, practice and programmes in this respect Identify employees exposed to noise and vibration Identify information issued to these employees Examine the safety equipment/clothing issued Test wearing rates Identify noise sources and attempts at reducing noise and vibration levels Examine whether noise and vibration levels have been reduced since the last safety audit (see figure 5.8) Examine noise assessment information (see figure 5.9) **Decision making environment:** From accident and dangerous occurrence data identify key decisions which have been made as a result of this information. Identify the decisions taken and note: * how the decision was taken * the time taken to take the decision * who took the decision * how the decision was implemented * what steps were taken to monitor the effectiveness of the decision Note weakness in hierarchical structures and delays in decision making on the safety audit noted sheets					

Figure 9.1 cont. A section from a safety audit checklist

	Quantifiable	Observed against quantifiable	Professional judgement	Satisfactory	Unsatisfactory
Bloggs Manufacturing Ltd Sheet 7 of 9 **Safety Audit Checklist** Carried out by: Date safety audit commenced: Date safety audit finished: Division/Department/Unit/Section/Individual: Safety audit of policy/procedures/practice/programmes/safety mix Location of safety audit: Signature of safety auditor: .... Date:					
Safety tasks					
Safety committees and representatives: Examine relevance of the Safety Representatives and Safety Committees Regulations, 1977. If applicable, examine all safety committee minutes since the previous safety audit in order to check: * the role of the committee within the decision making environment * membership * the importance given to the committee by the company * frequency of meetings * the role of the committee within the organisation * the effectiveness of the committee Identify any safety representatives within the company and: * examine their roles and tasks * examine their performance standards and objectives * examine their contribution to accident reduction targets * examine the frequency of their involvement in the decision making process					
Management involvement in health and safety: Examine the role of senior and middle management in the context of company health and safety objectives. Include all supervisors Note their support and enthusiasm and relate this to the individual accident and dangerous occurrence data relevant to their areas of responsibility Examine their roles and safety performance standards Note areas where involvement has been counter-productive to the overall safety performance standards					
Trade union liaison: Examine the role of trade unions within the context of company health and safety objectives Note their support and enthusiasm and relate this to the accident and dangerous occurrence data and relate these to their involvement and activities Examine their roles and safety performance standards Relate these to actual accident reduction targets since the last safety audit Note areas where involvement has been counter-productive to the overall safety performance standards					

Figure 9.1 cont. A section from a safety audit checklist

Bloggs Manufacturing Ltd Sheet 8 of 9 Safety Audit Checklist Carried out by: Date safety audit commenced: Date safety audit finished: Division/Department/Unit/Section/Individual: Safety audit of policy/procedures/practice/programmes/safety mix Location of safety audit: Signature of safety auditor: .... Date: .... **Safety tasks**	Quantifiable	Observed against quantifiable	Professional judgement	Satisfactory	Unsatisfactory
Industrial relations: Examine the role of the following in relation to the achievement of accident reduction/prevention targets: * legislation * collective bargaining * unilateral management decisions * unilateral trade union regulations * individual contracts of employment * custom and practice * arbitration * social conventions Note all areas where conflict has arisen and examine how the problem was resolved Note attitudes whether constructive or destructive Note liaison with other interested groups and note whether internal or external Note the assistance requested and given Examine how the information received was used or implemented **Disasters and emergency plans:** Any disasters which have occurred since the last safety audit should be covered in the preliminary stages of the safety audit. Here it would be necessary to obtain copies of any inquiry into the event whether internal or external. This detail can then be examined more fully within the context of this safety audit. List all recommendations and conclusions as these will be incorporated into the safety audit in the form of safety audit performance standards and objectives Where no disaster has occurred within the company since the last safety audit then: * does the company have a criteria for what it would call a disaster? * is there a policy for dealing with such an event? In which case it is necessary to: * examine all accident and dangerous occurrence data against the company criteria * examine the relevance and practicality of the disaster policy * examine evacuation plans and test where appropriate * examine disaster training programmes * do the workforce know and understand building evacuation procedures? Test the adequacy of the disaster plans					

Figure 9.1 cont. A section from a safety audit checklist

Bloggs Manufacturing Ltd Sheet 9 of 9 Safety Audit Checklist Carried out by: Date safety audit commenced: Date safety audit finished: Division/Department/Unit/Section/Individual: Safety audit of policy/procedures/practice/programmes/safety mix Location of safety audit: Signature of safety auditor: .... Date: .... <div align="center">**Safety tasks**</div>	Quantifiable	Observed against quantifiable	Professional judgement	Satisfactory	Unsatisfactory
Statements of health and safety policy: Examine the company health and safety policy document in accordance with figure 5.10 Examine: * how old or recent is the safety policy document * how is the information communicated to all employees * the languge used * the style of presentation * conforming to the policy statement by all employees Note all breaches of this safety policy document					

Safety audit notes:

Task	Comments

Figure 9.1 cont. A section from a safety audit checklist

Where local data is not available to help establish danger or to assist in the formation of an opinion, then it is quite normal to look at national or regional data sets. For example we know that coal dust can cause pneumoconiosis and there is sufficient national data to establish this link. This information is important when using professional judgement. It is not only medical issues which can be used in this way. We know also, that the faster the human being is asked to perform (or even drive his vehicle) the more likelihood is there of an accident, injury, death or dangerous occurrences. The evidence for this can be obtained from examining published statistics by such organisations as the HSE, the RoSPA, the Department of Trade and Industry (DTi) and the Department of Transport (DTp). Professional institutions and trade organisations are also a useful source of providing these statistics. Judgements should not normally be made without supporting information or evidence from some other source. Avoid passing opinions without any supporting evidence otherwise your opinion will be regarded as being no better than the next man's!

DESIGNING THE SAFETY AUDIT CHECKLIST

There is always a danger that commercially available safety audit documents do not quite fit the activities and tasks in all organisations. It is important, therefore, that management are aware of these limitations and that it is necessary to adapt or even design their own safety audit checklist.

It was stated at the beginning of this book that the safety audit should be conducted every two to five years because of the thoroughness of the activity. By all means carry out periodic safety reviews of policies, procedures, practices or programmes as these will assist the safety audit when it is finally carried out. There is no simple approach to carrying out a full safety audit and the documentation will need to consider all those headings provided on the summary checklists listed above. In addition, it will be necessary to itemise the various functions within these summary headings. A full safety audit, on average, can take around one month to complete but this of course varies with the size and nature of the business being audited. An example of one section of a safety audit checklist is given in Fig. 9.1.

Before designing the safety audit checklist it usually a good idea to have some idea of the order in which the safety audit is to be conducted. A good practice is to walk around the audit site(s) beforehand and note the main features to be examined against those which are more specific. Some general points to consider are:

- Obtain preliminary background information such as:
  - previous safety audit data;
  - accident records;
  - accident costings;
  - company statement on its health and safety policy.
- Obtain accident incidence and frequency rates (see below).
- Outline the aims and objectives of the safety audit.
- Outline the safety audit performance standards.
- Discuss the aims and objectives of the safety audit with senior management.
- Discuss the purpose of the safety audit with supervisors.
- Reassure all staff that the safety audit is to examine safety performance and is not to be regarded as something sinister!
- Request open access to all data and records.
- Identify all data access points.
- Do not announce the start date as the auditor/audit team will require to see things as they really are.
- Decide the order and format that the safety audit is to take.

From the basic information gathered at this stage of the safety audit, it is possible to draft the safety audit document and its layout. An example part of a full safety audit document is given in Chapter 11.

MEASURING SOME ASPECTS OF PERFORMANCE

It is said by some managers that if performance is not being measured then it is not being managed! This is equally so of safety performance and there are three levels to consider:

1. Organisational safety performance.
2. Process safety performance.
3. Job or task safety performance.

Safety performance is measured so that safety system performance can be monitored, controlled and improved at all three levels. Without such measures performance is not achieved (see Fig. 9.2) and managers have no basis for:

- specifically communicating safety performance standards to employees;
- knowing or understanding what is going on in the company;
- identifying gaps in safety performance which should have otherwise been identified and treated;
- providing appropriate feedback that is designed to compare performance to a standard;

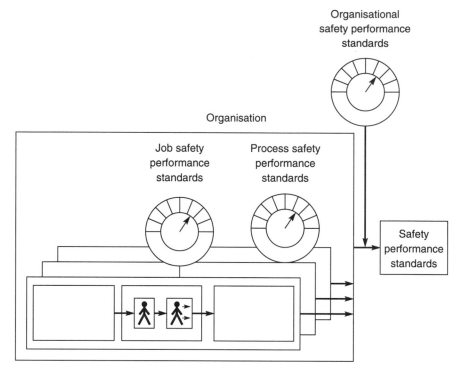

Figure 9.2 Measuring the three levels of safety performance within an organisational system

- identifying safety performance which could be rewarded;
- effectively making and supporting safety decisions regarding resources, procedures, practice, policies and programmes.

Without safety performance measures, employees at every level will have no basis for:

- knowing what is expected of them;
- monitoring their own safety performance and generating their own feedback;
- generating their own rewards and understanding what safety performance is required for rewards from others;
- identifying their own safety performance improvement areas.

However, merely establishing safety measures is not enough. There should be:

- sound measures that ensure that the right things are being monitored;
- a total measurement process and not just a collection of unrelated and counter-productive measures;
- a safety performance management system that converts the information provided by the measurement process into intelligent action.

In the general sense, performance means output and safety performance is this output quantified in terms of risk or danger to health and safety. The measures to consider then are:

- Identification of the most significant outputs of the company, departments, sections, teams and/or individuals.
- Identification of the 'critical dimensions' of performance for each of these outputs. Critical dimensions to:
  - *quality* include safety, accident prevention, accident reduction, ease of use, reliability, simple maintenance and appearance;
  - *productivity* include quantity, rate and timeliness;
  - *cost* include accident costs, labour, materials and overheads.

  Critical dimensions are derived from the output needs of the internal as well as the external customers within the organisation.
- Develop measures, objectives and standards for each critical dimension.

Measurement is the key feature in safety performance management and the following summarise this in relation to companies and human performance:

- Without safety performance measurement it is difficult to identify, describe and set priorities on safety problems.
- Without safety performance measurement, employees cannot fully understand what is required of them.
- Without safety performance measurement, employees will never know whether their performance is good or not.
- Without safety performance measurement there is no objective, equitable basis for rewards such as pay rises, safety performance bonuses, promotions or punishments such as disciplinary action, down-grading or even dismissal.
- Without safety performance measurement there are no triggers for performance improvement actions.
- Without safety performance standards management have to use a series of educated guesses.

These points should be considered by the safety auditor in cases where safety performance standards are either not available or where they are difficult to identify.

SOME SAFETY PERFORMANCE CALCULATIONS

At the initial stages of the safety audit planning and implementation stage it is necessary to obtain accident incidence and frequency rates which can provide a means of measuring safety performance over a period of time and then comparing this with accident statistics published by external sources such as the RoSPA, the DTp, the DTi and the HSE.

There are records available not only by employers but also the HSE and/or local authority. Reportable injuries under RIDDOR and these data, although collated by the HSE, are published by the HSC in their annual report and in a supplement to the *Employment Gazette.*

It must be remembered, however, that considerable under-reporting takes place within these statistics and it is only as accurate as employers make it within the terms of the current legislation. Some industries are worse than others for this. It is the responsibility of all organisations to follow the requirements of RIDDOR because without their full co-operation it is not possible for each industry accurately to compare its performance. Despite these shortcomings, incidence rates are useful in that they can be used to monitor safety performance between, say, divisions or departments over time.

In smaller organisations, such as those with less than 100 employees, it is probable that reportable accidents under RIDDOR represent only a small proportion of the number of minor injuries or dangerous occurrences which take place. This information can be converted into incidence rates so that management, and in particular the safety auditor, may monitor trends over a period of time (since the last safety audit) or between different parts of the operation. Calculating such incidence rates is based upon the HSE's own formula as follows:

$$\text{Incidence Rate} = \frac{\text{Total number of injuries in the period}}{\begin{array}{c}\text{Number of persons employed}\\ \text{during the same period}\end{array}} \times 100,000$$

It is accepted that this is rather a crude method of calculating the rate per 100,000 employees. There is also no allowance made for variations in part-time employment or overtime carried out. The formula would need to reflect the same periods of time if comparisons are to be made. Although incidence rates are useful to some organisations and safety auditors, some industries prefer to calculate accident frequency rates per million hours worked. This allows accident data to be put into perspective for comparison purposes. This particular method uses the hours worked rather than the number of employees in the formula thus avoiding distortions which may arise from the differences between full-time and part-time employee working hours. Frequency rates may be calculated during or over any time period as follows:

$$\text{Frequency Rate} = \frac{\text{Total number of injuries during the period}}{\begin{array}{c}\text{Total number of m/H worked}\\ \text{during the period}\end{array}} \times 1,000,000$$

Some safety auditors would rather use the following severity rate calculations as a means of comparing accident trends between appropriate departments, organisations and/or industries.

$$\text{Severity Rate} = \frac{\text{Total number of days lost}}{\text{Total number of m/H worked}} \times 1{,}000$$

The mean duration rate is used by some safety auditors which allows for a calculation to be made as a ratio between the number of days lost in relation to the number of man hours worked during the same period. The mean duration rate can be calculated thus:

$$\text{Mean Duration Rate} = \frac{\text{Total number of days lost}}{\text{Total number of accidents}}$$

A further duration rate can be calculated which expresses the number of man hours worked in relation to the number of accidents or injuries in the same period. The calculations can be made as follows:

$$\text{Duration Rate} = \frac{\text{Number of m/H worked}}{\text{Total number of accidents}}$$

These calculations are only a means of making simple comparisons between different sections, departments, organisations and industries. More reliable data and the use of rigorous statistical tests should be used at the end of the safety auditing process. The above should only be used at the beginning of the process in order to get a 'feel' for the situation before the safety audit gets under way more fully.

10 IMPLEMENTATION OF THE SAFETY AUDIT

In this chapter we shall look at the following:

- **Obtaining preliminary background information**
- **Outline the aims and objectives of the safety audit**
- **Outline the safety audit performance standards**
- **Discussions with senior management**
- **Discussions with supervisors about the purpose of the safety audit**
- **Reassure all staff about the safety audit**
- **Request open access to all data and records**
- **Identify all data access points**
- **Communication about the safety audit**

OBTAINING PRELIMINARY BACKGROUND INFORMATION

The implementation of the safety audit requires a number of issues to be considered so that all the preparatory data can be efficiently and effectively obtained. It is necessary to get hold of:

- the previous safety audit data;
- all accident and dangerous occurrence records;
- all accident costings;
- the company statement on its health and safety policy;
- any accident incidence and frequency rates which have been used by the company;
- details of any disasters which have occurred since the last safety audit;
- serious accident or disaster procedures approved by the company;
- any internal or external inquiry reports concerning any safety related matters within the company.

If there has been a previous safety audit conducted then this information will be an important starting point. It will be necessary here to identify the key task areas in the safety audit requiring action and this will need to form a part of the current auditing process. This is necessary in order to assess what action, if any, the company has undertaken as a result of the previous audit.

Discussion with the safety manager at this point is a useful place to begin as he should be in a position to provide the safety auditor with a summary of this information. The safety auditor will identify each recommendation from the previous audit and will examine what action, if any, has been taken to implement these recommendations. Any recommendations identified as not being implemented must form the basis of the discussions which will be held at the start of the safety audit with senior management.

Obtain all accident records and analyse the data into useful data sets. Gather dangerous occurrence data where available and analyse similarly for operational use. It may be appropriate to split accident frequency rates by location so that some comparative studies might be undertaken. At the same time, it could be beneficial at this stage to obtain the medical record summary sheets so that sickness and other absences might be assessed and used for the safety audit preparation.

The safety manager or company cost accountant should be able to provide a copy of all accident costing data which has been calculated since the last safety audit. These should be compared after allowing for inflation and other local variations in costing assessments. It will be necessary to calculate whether costs have gone up in real terms over the same period.

With regards to the company statement on health and safety policy, it is important to check whether it has been altered in the light of experience gained since the last safety audit. Where changes have been implemented it will be necessary to examine how and when these changes have been communicated to the workforce.

Company measures designed to cope with a disaster or other emergency should be gathered so that these may be checked and tested at the appropriate stage within the safety auditing process. It is also a good idea to obtain building evacuation times. These are usually given, for example, when fire drills are carried out and should be recorded on the fire procedures documentation. Check with the safety manager or local fire officer if in doubt. Evacuation timings should include the time taken to:

- clear the building(s) from the sounding of the alarm;
- arrive at the approved assembly points;
- carry out a register check of all personnel;
- report to the Senior Fire Officer with an accurate account of anybody who is missing.

It is also required that the safety auditor should familiarise himself with any reports whether internal or external which have been produced as the result of an inquiry into any major incident at the works since the previous audit. Any safety reviews which have taken place since then must also form an integral part of the safety auditing process. In such cases it is advisable to obtain the detailed accident investigation reports so that the contributory factors may either be confirmed or identified. Statements made by casualties, witnesses and other individuals must also be obtained and analysed.

OUTLINING THE AIMS AND OBJECTIVES OF THE SAFETY AUDIT

Before discussing the implementation of the safety audit with senior management it is important to outline the aims and objectives of the exercise clearly and concisely. In this case it will be necessary to define the mission statement. This could be:

'TO EXAMINE SAFETY PERFORMANCE WITHIN AN ORGANISATION'

or

'TO TEST SAFETY PERFORMANCE LEVELS AT ALL STAGES OF THE ORGANISATIONAL SAFETY SYSTEM'

Having decided on the main purpose of the safety auditing process it is normal to list the strategic objectives. These might be something in the order of:

- To examine safety performance at departmental, section and/or individual levels.
- To examine accident reduction and prevention strategies in the organisation.
- To examine the commitment of senior management to raise the profile of health and safety in the workplace.
- To examine safety policies, procedures, practices and programmes.

This list may be added to or changed depending on the nature of the industry being audited and the above is only given for general guidance.

Tactical objectives might be to employ:

- performance management theory and practice to carry out various tasks;
- role analysis techniques;
- task analysis measures;
- accident costing techniques;
- accident analysis procedures and practice;
- safety training measuring techniques;
- communication analysis programmes.

Again this list could be extended depending on the nature of the industry being audited. This gives some general headings on which tactical objectives might be set. These objectives will cover the actual techniques employed and would indicate the process that the safety audit was to take. In addition, the safety audit at an operational level would itemise those activities and tasks to be studied within the policies, procedure, practice and programme areas.

OUTLINING THE SAFETY AUDIT PERFORMANCE STANDARDS

This should be a statement of each activity and task illustrating wherever possible objective activities such as to:

- quantify all data wherever possible;
- minimise professional judgement and opinion;
- complete the preliminary study within five working days;
- complete (for example) site A, B and C within 14 working days;
- complete the safety audit in 30 working days;
- submit a preliminary report by June 15th;
- submit an executive summary and final report by June 30th;
- present recommendations and conclusions to the Board of Directors within seven working days of submission.

This list must be modified to suit each individual safety audit and is subject to the agreements made with senior management at the outset of the project. It might be necessary for some organisations to insist on a particular house style for the submission of reports and other presentations and this information should also be discussed with senior management. There would need to be checks with the safety auditor that agreed performance standards and goals were being achieved at specific stages of the safety audit. It would be appropriate to report back to senior management at agreed times as part of the performance appraisal stage of the audit.

DISCUSSIONS WITH SENIOR MANAGEMENT

An important aspect of the safety audit implementation process is the discussions with senior management about the purpose of the safety audit. It is best to speak to each individual departmental head before speaking to them as a group. In this way individual worries and concerns can be dealt with in a more confidential manner. When all individual heads of department have been seen then a more collective approach to the necessary discussions can be held regarding the general issues to be considered under the safety audit programme. It would be at this stage in the process that confidential discussions would concern previous safety audit recommendations and their implementation. At the same time it is useful to discuss the question of data and information access. It will be at the individual discussion stage that discussions would take place regarding the procedures to be adopted within each area of responsibility. Also, it would be useful to obtain the names and locations of all supervisors so that permission can be obtained in order to discuss more operational matters concerning the safety audit. At a more collective level the discussions would include:

- details of performance standards;
- agreement of performance standards;
- performance appraisal meetings with individuals;
- agreement on free access to data;
- house styles for the submission of reports;
- the sequence that the safety audit will take;
- general comments concerning the collection and use of data from individual departments.

At this stage, it would be appropriate to discuss and to agree the start and finish dates so that these can form a part of the performance standards and to give a general indication and direction as to how the safety audit will progress.

DISCUSSIONS WITH SUPERVISORY STAFF

Having had the opportunity to discuss the strategic and tactical issues concerning the safety audit with senior management it is necessary to provide an opportunity to discuss the operational issues with each supervisor. This allows for a two-way exchange of ideas concerning individual areas of responsibility to take place. Some supervisors may never have experienced a safety audit before, therefore it is important to spell out in plain language (jargon free) what the aims and objectives of the process are. There are some supervisors who will also be unfamiliar with safety performance standards and how they are to be assessed within the safety audit. These should be explained so that no confusion or misunderstandings can occur.

It would be diplomatic at this stage to seek the approval of supervisors to examine their safety records, tasks and roles within the organisation and to speak with and/or observe sections, teams and individuals within their area of responsibility and to:

- discuss the safety audit with all line managers;
- outline safety audit aims and objectives to line managers and foremen;
- to give the start and finish dates;
- give details of the safety audit performance standards where the information is directly relevant to the line manager or supervisor concerned;

- answer any questions or concerns that individuals might have about the process to be undertaken and the part that they have to play within it.

It is very important to have the support of all managers and supervisors involved directly and indirectly with the safety audit and to explain the need to reduce and to prevent accidents. It will also be important to identify at the preliminary stages whether any department, unit/team has significantly higher accident rates than others as this will influence the discussions with those individuals involved. Clearly, where a sub-unit within the organisation has a particularly unsatisfactory accident record, then this will influence the discussions and the aims and objectives for that particular area of activity. This too will be given special attention within the safety auditing process.

REASSURING STAFF

Employees can be highly suspicious of people walking around the works gathering or recording information if they are unaware what such an examination is about. It is essential, therefore, to explain, or have explained by senior management and supervisors, details of the safety audit and what it is intended to achieve. Supervisors can be issued with a safety audit brief prepared by the safety auditor and this would explain in a clear and concise way the aims and objectives of the auditing process. Such a brief would also include start and finish times of the safety audit and would promise a debriefing exercise carried out at a later date by the supervisor concerned. It is necessary to inform staff of the auditing process in order to:

- avoid any misunderstandings about the safety audit;
- emphasise at the outset that employees should do normal things during the audit in order to avoid 'play acting' or doing non-normal things;
- gain their support and confidence;
- provide an opportunity to develop a good rapport with employees generally;
- produce the finally agreed safety audit implementation plan;
- introduce new concepts or provide explanations of certain aspects of the safety audit which may be identified as a cause of worry to employees;
- outline the importance and commitment of senior management to the safety auditing process;
- provide an opportunity to give the supervisors' and trade union viewpoints.

It will also be useful to inform safety representatives and safety committees at this stage. If they have done their job properly, they will have assisted management and in particular, the safety auditor in this task by disseminating early information to all employees about the safety audit and its purpose.

Any areas identified as the cause of staff apprehension or uncertainty should be dealt with here. However, staff should be aware that the safety audit is a rigorous examination of all procedures, practice, policy and programmes within the working environment and it must be able to make a critical examination of all facets of these four features. Such an examination will look closely at employee:

- attitudes
- behaviours
- skills
- knowledge
- abilities
- roles
- tasks
- performance standards
- performance appraisals
- achievements

within the context of accident prevention and reduction within the workplace since the last safety audit.

IDENTIFYING ALL ACCIDENT AND OTHER SAFETY AUDIT DATA POINTS

It is important to know what safety audit data is required in order to identify the appropriate data collection points. Once the level and type of data has been agreed it is necessary to draw up a list as to where this data is to be obtained from. For example, if it was agreed that accident data was required then it might be established that this is usually kept locally by each supervisor, first-aid post, personnel officer or by the safety manager. Likewise, if medical records were needed then these might be obtained from the company medical practitioner, nurse, first-aid post, personnel or safety departments. Other information required by the safety auditor concerns roles, tasks, objectives, performance standards and assessments, therefore, all data points relevant to these areas need locating and the data format examined. Where data is kept in a format difficult to use for safety auditing purposes then this fact will form a part of the safety auditing process and should be commented upon at the appropriate time. Discrepancies in data or quality must also form a part of the safety auditing process.

Where the safety auditor is independent from the organisation being examined it is usual to speak with the following key personnel as part of the identification of safety audit data points within the implementation phase:

- Safety manager.
- Personnel manager.
- Adminstration or Office manager.
- Company accountant.
- Senior medical or nursing staff.
- Departmental heads.

This will confirm the level and type of information currently kept and where it is located within the organisation and who controls it from a quality perspective on a daily basis. It will also provide an opportunity to discuss not only the format but the reliability of the data. This will also indicate the way in which the safety audit will proceed from this point.

GAINING OPEN ACCESS TO ALL DATA

This must be agreed by senior management during the preliminary stages of the safety auditing implementation process. Continuing with a safety audit will be a fruitless exercise unless full access is given to all data as it is required. Confidentiality of the information received and used is naturally assured and it is for internal use only. This aspect should be covered by professional codes of practice such as those covered by organisations such as the Institution of Occupational Safety and Health (IOSH), the International Institute of Risk and Safety Management (IIRSM) or the Royal Society of Health (RSH). It is essential, therefore, that properly qualified and registered staff are used to conduct safety audits. This assures not only such areas as the confidentiality of data used but also provides confirmation of competence and professionalism. This is particularly important where the safety auditor is not an employee of the organisation being audited.

Managers must be visible in the workplace for the promotion of safety as well as for production. In this way, it is possible to become a safe production unit.

In the more complex industrial operations such as mining, there are often several facilities tied to an administrative or headquarters-type location. In such cases, individual sites sometimes have little or no control over areas of programmes or systems governing their operations, particularly those set up for the whole organisation. Examples might include the safety programme outline, the disciplinary programme or the needs of the workforce. It is important, therefore, that locations are evaluated for those systems which they control and that headquarters type centres are analysed for those that they command. The following is a summary checklist for the type of information for which access points must be identified:

Headquarters or group data sources

Obtain the loss control programme data for:
- employees;
- processes;
- contractors;
- equipment.

Obtain data sources for:

- the roles of senior managers in the accident prevention programme;
- whom the safety manager is responsible to;
- how often the accident prevention programme is reviewed (not audited) internally for effectiveness;
- if there is a programme for training supervisors based on established need;
- if senior management occasionally attend site safety meetings;
- if there is a system which recognises supervisors' and employees' safety achievement;
- whether aims and objectives are required for safety performance;
- whether all objectives are realistic and measurable;
- how and when the safety objectives are measured;
- the consideration which is given to matching man, machine, and the environment in the hiring and placement of employees;
- obtaining the programme data for the selection and purchase of:
  - facilities;
  - equipment processes;
  - protective clothing and equipment.

Site or satellite location data sources

Obtain the accident prevention programme data for:
- employees;
- processes;
- contractors;
- equipment.

Obtain data sources for identifying:

- the role of management in the safety programme;
- whether management recognise supervisors and employees for their safety achievement;
- whether management initiate procedural and motivational communications to staff;
- whether management participate in accident prevention inspections of the facility;

- whether management attend safety meetings with the workforce;
- whether safety aims and objectives are set jointly by the entire management team at that location;
- whether the safety objectives are measured;
- whether senior management review all major incidents and subsequent statistical analyses;
- whether there is a preventive maintenance programme;
- whether there is a system of selection and placement of employees;
- whether there is a job analysis programme in operation;
- the safety training programme for: management development and supervisory safety task training;
- whether there is a safety co-ordinator and what his responsibilities are.

Obtaining accident and dangerous occurrence data

Obtain data sources for identifying whether:

- remedial measures attack the factors contributing to the incident happening;
- the safety system is used to follow up on preventive measures;
- the accident investigations show the potential for severity;
- all fatalities, major injuries and illnesses, and high cost incidents are investigated;
- departmental heads review all incidents in their departments;
- safety departments review and follow up all incident reports;
- thorough analyses are made of all incidents for trends and if these analyses are reviewed by senior management;
- there is a method of determining the number and type of incidents resulting in lost time from off-the-job injuries and illnesses;
- there are off-the-job incidents analysed for type and costs to the facility and how these analyses are communicated to senior management;
- employees aware of safety programmes which include off-the-job safety;
- what off-the-job safety information provided to the families of employees, if any.

COMMUNICATIONS REGARDING THE SAFETY AUDIT

It is necessary to disseminate information concerning the safety audit to the general workforce and to do this it is important to set objectives in the way described earlier in this book (see Chapters 2 and 6) and to decide on the appropriate messages to use and the most appropriate vehicles to carry these messages. Some messages might include:

- What the purpose of the safety audit is.
- The form that the safety audit will take.
- Details of safety performance standards and why and how they will be examined.
- Organisational, departmental, section/unit/individual aims and objectives.
- Why and how safety performance assessments will be examined.

Appropriate vehicles to inform employees might be a:

- leaflet;
- poster;
- letter;
- talk to groups or individuals;
- mixture of all or some of these.

In Chapter 6, it was found that listening is our main communication ability. We know that over 70 per cent of our time awake is spent in some kind of communication activity, and this was summarised as:

- writing   9 per cent
- reading   16 per cent
- speaking   30 per cent
- listening   45 per cent

Although, in terms of time spent, listening appears to be the most important of our communication abilities, and writing the least, it is interesting to look at our educational system to see the priority that is given to the learning of the different communication skills. Do our safety communication priorities involve:

Writing   is this the most common method being audited?
Reading   is this the next most popular means?
Speaking   how often is this used?
Listening   do employees listen to you effectively?

This listing must be compiled by the safety audit team after an examination of the communication methods used in the organisation, departments, sections and to individuals. The amount of formal training given to these skills should also be examined in order to compare this ordering with what actually happens. Has a rational training programme been introduced since the previous safety audit which gives a priority to the skills that were most important such as:

- Developing our listening habits because listening is the first communication skill we develop, and the least taught.
- The value of good listening, and the problems created by poor listening are now well known. Recent surveys of safety managers have shown that listening, and listening related skills, are cited as the most critical managerial skill, and the one in which training is most needed.

- We know how important good listening is from our own experience of life. The problem is to identify the skills and techniques of good listeners.

Some managers argue that it is not necessary to disseminate information to the employees beforehand as this might influence what is observed. It should be remembered that the safety auditor is not always too interested in the observed as he will only use this as supplementary information to support the hard data which he will gather as part of the safety auditing process anyway. Surely, to disseminate information at the start of the exercise will place health and safety squarely in people's minds and will not detract from the value of the safety audit or influence any decisions which might be made as a result of it. Accident and dangerous occurrence data will help to confirm this.

Some organisations which have to operate in high risk industrial settings such as the chemical, petroleum or nuclear industries may wish to extend their communications externally. This might be a useful public relations exercise and one which could easily reassure the local community or local authority. An organisation which is seen to adopt a sound self-assessment policy of its own safety performance is one which will build upon local confidence and trust. This aspect should be considered at the same time as internal communications are planned.

11 COMPLETING THE SAFETY AUDIT DOCUMENT

In this chapter we shall look at the following:
- **Gathering the information required**
- **Completing the documentation**
- **Monitoring the safety audit**

GATHERING THE INFORMATION REQUIRED

This is the most fundamental part of the safety auditing process and requires a number of approaches such as:

- the formal interviewing of employees;
- questioning;
- an examination of records, documents and statistics;
- observation;
- reading internal and external reports, letters and memoranda;
- talking and listening casually to employees.

Formal interviews with employees

These should be prepared as the safety audit progresses and it would be required to speak formally with:

- the very senior managers in the headquarters organisation;
- the safety manager;
- with the safety representatives;
- with the Chairman and Secretary of the Safety Committee or Advisory Group;
- any regional or divisional heads;
- departmental heads;
- supervisors;
- casualties;
- those who have reported two or more dangerous occurrences since the last safety audit.

Although these might be seen as formal events it is vital that the safety auditor prepares the structure of the interview and follows this format throughout the interview. Such interviews should not normally need to be longer than 20 minutes. It might be appropriate to record the meetings or to seek permission for the interview to be recorded so that appropriate records can be compiled concerning the specific areas of the interview. Items that you might wish to discuss on these occasions might be:

- accidents frequencies within specific areas of responsibility;
- accident costs;
- some views of the person being interviewed about their involvement in accidents, accident investigation or accident analysis;
- some views and maybe some evidence on the contribution that the person being interviewed has made to accident reduction and prevention since the last safety audit;
- what actions have not been taken that, in the opinion of the person being interviewed, should have been in order to reduce or prevent accident or dangerous occurrences from happening within their areas of responsibility.

The list should not be too long. It should be structured in such a way that the safety auditor can obtain all the information he would need (based upon preliminary work as outlined in Chapter 10) to ask within the shortest possible time scale.

Questioning

This can take place without the need to set the meeting within a formal framework. Questioning tends to be interpreted as a formal means of obtaining information but not necessarily within a formal environment. Questioning can take place at any location within the working environment and can last for any length of time. However, it is usual to structure the questioning process so that a logical format can be adhered to during the exercise. It might be necessary to question:

- accident casualties;
- witnesses;
- participants in dangerous occurrences;
- members of the safety committee or safety advisory group;
- accident investigators;
- operators in high risk activities;
- maintenance personnel;
- administrators.

It should be remembered that the objective of questioning is to obtain information that will help confirm quantifiable or objective data or provide a supplementary base on which to assist the safety auditor's decision-making process. This is particularly important where subjective data is concerned and the safety auditor needs to formulate an opinion based upon all available data.

Records, documents and statistics

All records, documents and statistics held in the organisation which relate to health and safety should be available for scrutiny by the safety audit team. Examples could include:

- medical records;
- accident and dangerous occurrence records and data;
- accident costing data;
- all maintenance and servicing schedules;
- safety equipment records;
- safety equipment issue details;
- first-aid treatment records;
- records relating to fires;
- COSHH records;
- task analysis records;
- safety performance records;
- safety performance appraisal documentation;
- role analysis documents;
- minutes of safety committee meetings.

This list should be developed at the preliminary stages of the safety audit implementation phase so that the safety audit team know exactly what information is required and the locations where the data is held. Again, objective or quantifiable data is the most valuable whilst other supplementary information is used to assist in helping the safety audit team to firm up on professional judgements or opinion.

Observation

There will be some activities or tasks within the safety audit which will require observations to be made. Data might suggest that these be:

- operational tasks;
- operational roles;
- methodologies;
- safety systems in operation;
- maintenance operations.

This list should also be drawn up at the preliminary stages so that high risk tasks, roles, working procedures and practice can be observed at first hand and appropriate plans made within the safety audit to do this.

Where certain tasks or operational roles are to be observed it is necessary to have details of the task analysis/role analysis documentation so that each component of the job can be clearly identified. Again, the emphasis is on obtaining objective data or supplementary data which can be used to help formulate or consolidate professional judgement or opinion.

Internal, external reports, letters and memoranda

Internal and external reports are a rich source of information. Where appropriate it might be necessary to see certain letters or memoranda. These may be confidential so professional codes of conduct apply. At this stage it is important to consider not only the problems which caused the origination of the internal/external reports but also to try and identify what action has been taken. Where external reports have been obtained it might be necessary to talk with the authors of such reports in order to clarify any points or to expand on appropriate concerns. In terms of letters and memoranda, all relevant documents since the last safety audit should be obtained. As is the case in all communications activities it will be necessary for the safety auditor to note the times taken in responding to any matter which required swift and prompt action in the cause of health and safety. Where unacceptable delays can be identified these issues should be recorded as an integral part of the safety audit and will form not only the basis of the safety audit report but a part of the safety audit debriefing process which will come at the end of the safety audit.

Casual conversations with employees

These can be the most productive and lucrative sources of safety information in the workplace. It is important to build up a rapport with the workforce during the safety audit both so that they can feel more at ease and also so that they can feel free to ask questions about any doubts which they might have. This is particularly important in the area of substantiating and enlarging on the dangerous occurrence data which will already have been obtained. A great number of employees are reluctant to report dangerous occurrences because they either feel stupid at having allowed a potentially hazardous situation to arise or for the fear of disciplinary action if they report them. If the workforce feel at ease with the safety audit team and that anonymity is assured casual conversations can be most revealing. Where the reporting of dangerous occurrences is very small (less than 10 in a company of 500 employees or more since the last safety audit) it is important for the safety auditor to examine the environment in which dangerous occurrences are reported.

Companies with bad dangerous occurrence reporting information may either not encourage the information to be reported officially by employees or the system is such that it is incapable of realising the information anyway. Honesty boxes may be a simpler solution or programmes of reward bonuses to employees who encounter a dangerous occurrence and can present sound recommendations for preventing the occurrence from happening again.

COMPLETING THE DOCUMENTATION

When completing the safety audit documentation it is important to keep the data gathered into appropriate groupings for:

- headquarters;
- regions and/or divisions;
- departments;
- units/sections/teams;
- individuals.

Also, within these categories will be subheadings for the safety policies, procedures, practices and programmes which are to be examined against those which were examined at the last safety audit. Additionally, a summary recording of event classifications such as will be necessary:

- satisfactory;
- unsatisfactory;
- objective;
- subjective;
- professional judgement.

Where something is said to be satisfactory then the written objective or performance standard has been met. Something unsatisfactory means that the goals have not been achieved. These are simple statements and should be clearly interpreted as to whether the standard has been achieved or it has not.

Objective matters relate to the safety audit data and it is only possible to record something as satisfactory with any degree of certainty where the information being used to decide this matter is objective in nature. For example, the performance standard in the tool room was to reduce accidents by 2 per cent. Having checked the tool room accident records it will be possible to see whether this is the case or not. A reduction of 1 per cent would be below the written performance standard and would in this case be recorded as unsatisfactory. There could be several reasons why a performance standard has not been met such as:

- incorrect performance standards not being set initially;
- no performance appraisals;
- poor performance appraisal techniques;
- inappropriate or no supervision;
- poor task and role analysis;
- unrecognised socio-technical conflict;
- poor safety policies, procedures, practice and programmes.

The safety auditor would examine why a performance standard was not achieved and would make appropriate conclusions and recommendations at the end of the safety auditing process.

Subjective issues are those which cannot be quantified. In these cases it is always appropriate to get supplementary information in order to help substantiate or confirm an opinion. For example, a safety performance standard might exist to provide safety education to the workforce in respect of improving certain safe skills when using particular tools. Whilst a properly conducted scientific audit would need to be undertaken to try to quantify the success of this it would be virtually impossible to discover whether it was the safety education programme alone which contributed to improved skills and thus accident reduction/prevention or whether it was from some other source and in what proportions? Subjective safety performance standards should be avoided.

Professional judgement is an area where an opinion is required as to whether the performance standard has been met or not. For example, an organisation may have a performance standard which states that all fire extinguishers will be maintained in accordance with manufacturers' recommendations. But will they actually work when needed (a reliability issue)? Maintenance schedules could be examined and if necessary the safety auditor could request the fire equipment servicing people to actually try one or two out. Although there is no guarantee that the replacement ones will actually work when required the supplementary activity will help support the professional judgement to be made on this issue.

MONITORING THE SAFETY AUDIT

When examining safety audit approaches for control purposes, the safety auditor needs to focus on the sets of objectives that have been recognised as the formal reporting system and which will provide the main basis for considering any rewards or sanctions. These activities are seen as the primary triggers for organisational pressure and intervention if safety targets are missed or not met. The main purpose of the safety auditing process, therefore, is to help monitor safety performance within the organisation. Likewise, it is necessary from time to time to monitor the progress of the safety audit itself. This is necessary because there will have been safety performance standards set and objectives agreed at the outset of the safety audit. It would be wrong to proceed through the entire safety auditing process without it undergoing monitoring at various phases. Safety audit performance appraisal is an essential feature of the actual auditing process and requires the safety auditor to:

- identify specific problems encountered which may not have been previously been anticipated;
- discuss any problems with appropriate managers and supervisors;
- check progress;
- make any changes which might be necessary and to re-adjust any additional performance standards;
- confirm that safety auditing resources are being utilised efficiently and effectively;
- distinguish more accurately between statutory and voluntary monitoring requirements;
- discuss aspects of the safety audit with the organisation.

The number of safety audit performance appraisals would depend on the size of the organisation being audited, the type of industry involved and the projected safety audit timing.

Unexpected problems can sometimes occur which have the potential to delay the safety auditing plan. Absences by key staff, mislaid files and other irregular administrative malfunctions are a constant source of worry for the safety auditor. Such matters need to be discussed with appropriate senior managers when the delays or problems are likely to affect the safety audit process. A safety audit performance standard would normally include the immediate discussion with senior management when a problem is encountered. Safety audit performance appraisals would, therefore, examine such matters. Any changes to the auditing process will require revised performance standards to be agreed. Whilst progress will form a part of the safety audit monitoring process there will be an opportunity provided to discuss specific aspects of the safety audit and to comment on more general issues. Managers should ensure that safety auditors or safety audit teams are given sufficient freedom and authority in the organisation to allow them to do their jobs efficiently and effectively without having to constantly refer back to management.

Managers also play an important role within the safety auditing process. They must ensure, wherever possible, that all areas of potential delay are monitored separately by management in order to allow a smooth passage for the safety audit team. On the other hand, the safety auditor will try to keep any disruption in the working areas to a minimum. Managers must operate an orderly and regular review of their own objectives and the implementation of these during the safety audit. The monitoring of specific objectives at critical periods is also important as it is for the safety audit team. Any fluctuating conditions might influence the safety auditing process and if these issues are allowed to go unnoticed or unchecked then problems inevitably arise.

12 EVALUATING THE SAFETY AUDIT

In this chapter we shall look at the following:

- **Personal evaluation**
- **Group/team evaluation**
- **Section evaluation**
- **Department evaluation**
- **Organisational evaluation**
- **The analysis process**
- **Notes on performance-related pay**
- **Writing the safety audit report**

Post safety audit questionnaires to individuals, supervisors and senior managers are a means of providing essential feedback into the safety auditing system.

PERSONAL EVALUATION

Managers must not only constantly evaluate themselves, they must also encourage personal evaluation amongst the workforce. Such an activity should be a clearly defined objective and should form part of the safety performance standards which are set with each individual. Self-monitoring and evaluation is referred to as skills inventories. An example of a safety skills inventory is given in Fig. 12.1.

From these questions the safety auditor can examine the mix of an employee's abilities. Managers will collect this information either by direct interview/questioning, by telephone contact or by observation. Such information may be sent to management periodically as part of the safety performance standards and supervisors may wish to add their own comments. It is normal practice today to computerise this information in order to match talents, skills and abilities. This is a valuable source of personnel information because it can also be used to match skills and abilities to new openings and promotions within the organisation.

GROUP/TEAM EVALUATION

The evaluation of groups or teams follows a similar process to that described above for the individual. A typical group/team safety skills inventory is provided in Fig. 12.2.

Bloggs Manufacturing Ltd			Page 1 of 2

Safety Skills Inventory

This part of the form is to be completed by the Safety Manager.

No. of accidents since the last safety audit

Date inventory valid from: ▭

Name of employee: ▭   Employee ref no: ▭

Job title: ▭   Experience: ▭

Age: ▭   Number of years employed by this company: ▭

Other jobs held in this company:

Job title	From	To
Job title	From	To
Job title	From	To
Job title	From	To

Jobs held with other companies:

Job title	From	To
Job title	From	To
Job title	From	To
Job title	From	To

This part of the form is to be completed by the employee

Special skills: List below any skills which you have even if they are not used on your present job.

Include types and names of machines or tools with which you have experience.

Skills	Machines	Tools
Skills	Machines	Tools
Skills	Machines	Tools
Skills	Machines	Tools

Briefly describe your present duties:

Figure 12.1 An example of a skills inventory form

Bloggs Manufacturing Ltd →   **Page 2 of 2**

Safety Skills Inventory

No. of dangerous occurences reported
since the last safety
audit

Describe your responsibilities for health and safety:

Who is your supervisor?

What is your education?

List the safety training courses you have attended since the last safety audit

Special safety training received and qualifications gained

This part should be completed by the Safety Manager:

Overall evaluation of this employee's safety performance

List safety deficiencies

Summarise action to be taken to resolve safety deficiencies

Signature of Safety Manager   Date

Figure 12.1 cont. An example of a skills inventory form

Bloggs Manufcturing Ltd

Safety Skills Inventory - Groups or Teams

This part of the form is to be completed by the Safety Manager.

No. of accidents since the last safety audit

Date inventory valid from:

Name of employees: Employee ref nos:

Group title:

Ages: Number of years employed by this company:

Other jobs held in this company:
Attached individual safety skills inventories for each team member

List those accidents or dangerous occurrences the team have experienced since the last safety audit

List safety training courses attended by each group member since the last safety audit

List specialist safety training received and qualifications gained

Signature of Safety Manager   Date

Figure 12.2 An example of a skills inventory form for groups or teams

The method of gathering this information will also be summarised as a performance standard and will be subjected to the safety performance appraisal process. Members of groups of teams (e.g. maintenance teams) will also have individual skills inventories. This information will assist management to create and build more cohesive groups and will contribute to greater efficiency and safety effectiveness.

SECTION EVALUATION

Sections may be comprised of groups or teams and individuals within the same working environment. Performance standards would include individual skill inventories and those of the teams within the section. The section itself would have safety objectives and safety performance standards and these would form a part of those agreed with the section leader and would be subject to safety audit. An example of a section evaluation checklist is given in Fig. 12.3.

Managers will need to have this information ready for the safety audit team to examine together with the group, team and individual skill inventories.

DEPARTMENTAL EVALUATION

When auditing the department, the safety auditor will require objectives and inventory skills from the following:

- sections;
- groups;
- teams;
- individuals.

In addition, the safety auditor (or in the larger organisation safety audit team) will need to examine safety performance standards from each of these areas in turn and those which relate to safety objectives and safety performance standards of the department itself will then be compared and assessed. The purpose is to review the department's work. Fig. 12.4 lists the major areas to be covered. Here the safety audit team should:

- identify who is responsible for each activity in each of the groups (including individuals);
- determine the objectives for each activity;
- determine each set of safety performance standards;

Bloggs Manufacturing Ltd

Safety Skills Inventory - Sections

This part of the form is to be completed by the Safety Manager.

No. of accidents since the last safety audit

Date inventory valid from:

Name of employees: Enlarge this where necessary

Employee ref nos:

Section title:

Section experience:

Other jobs held in this company:
Attach individual safety skills inventories for each section member

List those accidents or dangerous occurrences the section have experienced since the last safety audit

List safety training courses attended by section members since the last safety audit

List specialist safety training received and qualifications gained

Signature of Safety Manager   Date

Figure 12.3 An example of a skills inventory form for sections

Topic	Completed
Number of accidents	
Number of dangerous occurrences	
Deficiencies	
Affirmative action goals	
Progress towards safety goals	
Safety performance standards	
Safety performance appraisals	
Roles	
Tasks	
Safety skills inventories	
Safety training	
Safety learning rates	
Safety training objectives	
Safety orientation programmes	
Compliance with the law and other regulations	
Standards for measuring safety performance	
Safety communications	
Discipline procedures	
Change and development procedures	
Management rights	
Employee rights	
Safety committees	
Safety representatives	
Safety policies	
Safety practices	
Safety procedures	
Safety programmes	
Employee feedback on health and safety issues	

Figure 12.4 Departmental safety audit checklist

- review the policies, procedures, practices and programmes in the department;
- examine all records held in the department;
- prepare a report commending safety performance achievements;
- develop an action plan to correct errors and under-achievement;
- ensure management follow-up on the action plan.

Within the departmental safety audit it is appropriate to examine how well managers comply with health and safety policies, practices, procedures and programmes and to see how well safety performance standards are appraised by senior management. Besides ensuring compliance, the safety audit can not only improve the department's accident prevention and reduction objectives but can also improve departmental image. Also, research has shown that operating managers gain a higher respect for the department when their views are actually sought particularly where subjective issues are concerned (e.g. employee satisfaction). If these views are acted upon then the department will be seen as being more responsive to their needs. These actions will improve the contribution to overall organisational objectives.

Effective departments meet both company safety objectives and employee needs. When employee safety needs are not met then staff turnover, absenteeism and increased union activity are more likely.

ORGANISATIONAL EVALUATION

This is the last area to be safety audited and can only take place after:

- each region or division (where appropriate) has been safety audited;
- departmental safety audit information has been gathered;
- section data has been received;
- group or team skill inventory information has been examined;
- individual safety performance appraisal data has been analysed.

Now it will be possible to examine, for example, whether the company statement on health and safety policy is applied or not and how relevant it is to the needs of the organisation in terms of accident prevention and reduction achievements. The safety auditing management system is illustrated very simply in Fig. 12.5.

THE ANALYSIS PROCESS

Having gathered all the data required for the safety audit it is necessary to look at some of the analyses which will have to be undertaken in order to interpret some of the information gathered. In Fig. 12.6 a summary of the approaches to safety audit analysis is given.

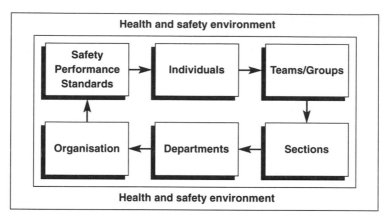

Figure 12.5 A simple safety audit management system

Comparative analysis

This is the most simple to undertake and it will require the safety auditor to take another group, section or department within the organisation as a model. Results are then compared. Such analysis is valuable when auditing for the first time a safety programme such as an alcohol and drug rehabilitation exercise or a stress management programme where comparison of one department with another is useful. Here also, the safety auditor, if an external member of the organisation, may be able to provide a benchmark for the comparative study. Standards set by the safety auditor or safety audit team may be used for comparison purposes. This is particularly important when considering subjective issues or when placing objective data into perspective.

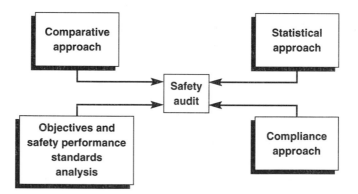

Figure 12.6 Approaches to safety audit analysis

Statistical analysis

These are essential tools for the safety auditor and statistical measures of performance can be based upon the data which has been gathered as a part of the safety auditing process. Let us assume that, for example, that information gathered as a part of the safety audit indicated that as a result of a high accident rate staff turnover and absenteeism was suspected to be higher than since the last safety audit. A statistical approach is usually supplemented with other analyses and other data sources. If our company employed (on average) 250 employees during the month of the safety audit and 10 employees quit then the staff turnover rate can be calculated thus:

$$\frac{\text{Number of employees quitting}}{\text{Average number of employees}} \times \frac{10}{250} \times 100 = 4\%$$

Staff turnover is said to be at 4 per cent and may now be compared with previous occasions of significance. If, however, during the safety audit 15 were absent on a particular day then:

$$\frac{\text{Number of employees absent}}{\text{Number scheduled to work}} \times \frac{15}{250} \times 100 = 6\%$$

Absenteeism is said to be at 6 per cent and can likewise be compared with other occasions of importance or significance.

Other statistical applications can also now be applied to data as outlined in Chapters 2 and 5 of the Handbook of Safety Management.

Compliance analysis

This is a basic feature of the safety audit process and examines whether safety policies, programmes, procedures and practice comply with legal requirements. The safety audit will examine all the features of the auditing process from a compliance standpoint. Compliance analysis also includes examinations of:

- safety performance standards;
- safety performance appraisals;
- administrative procedures;
- discipline;
- compensation.

It is essential that managers and employees know that they are to be examined from time to time in order to test whether they are complying with internal rules and legal requirements.

Objectives and safety performance standards analysis

This concerns the examination of objectives and goals and would also include safety performance standards and appraisals. Specific goals which can be measured and, where appropriate, compared with previous occasions or other comparable data. None of the above approaches can be used in isolation. Safety audit teams will usually use some or all of these approaches to analysing the safety audit data. Actual strategies employed will depend on the specific activities under evaluation. Fig. 12.7 suggests how the safety audit team provides feedback on those activities examined as part of the safety auditing process.

Unfavourable feedback must lead to corrective action being taken which will improve the contribution to accident reduction and prevention within the organisation.

NOTES ON PERFORMANCE PAY

Managers very often have to act in the interests of the shareholders but only if they have the right incentives. The way that managers and employees are measured and rewarded must be considered within the capital investment processes. Companies have formal procedures for evaluating the performance of their capital investments and there are three aspects to performance measurement. These are:

1. Companies need to monitor projects and schemes under construction in order to ensure that there are no serious delays or cost over-runs.
2. Companies generally conduct a post-audit which help identify problems which require remedial action.
3. There is on-going performance measurement which is conducted via the company accounts and control systems.

It is important, therefore, that safety schemes under construction should be subjected to the above criteria but there can sometimes be conflict between an organisation's objective to reward performance when compared directly with safety performance. There is a wealth of evidence now available that shows that some 'incentive' schemes can be counter-productive and can be a significant contributory factor in accidents. For example, road safety experts are aware of a correlation between delivery performance standards and accidents. One company, for example, employs delivery drivers whose bonus is calculated on the number of deliveries which can be made in a specified period of time. This scheme encourages drivers to speed and drive without due care and attention as the drivers' records and accident figures illustrates.

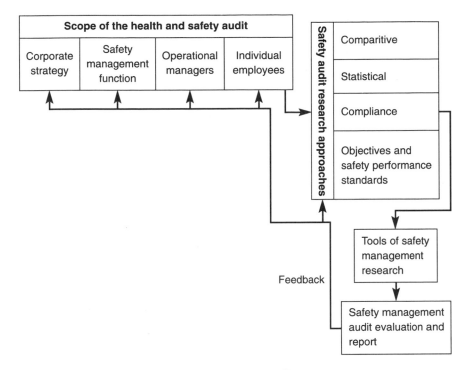

Figure 12.7 An overview of the safety management audit process

This scheme was shown to be counter-productive and not in line with the aims and objectives of the scheme when it was first proposed. A financial post-audit failed to pick up this point but the safety audit did. It was found that the number of road accidents and the cost to the organisation of these events made the scheme too expensive to continue. The scheme was replaced with another which regarded safety performance rather than other aspects of performance such as speed.

In another company, workers were paid on the number of items they could make in any one hour. Items produced above a certain number formed part of the bonus scheme. Because speed was again considered as the prime measure of performance the safety audit team not only discovered an unacceptably high reject rate amongst the items being produced but also found that the accident rate was significantly higher than the period before the scheme was introduced. The scheme was changed to one where both quality and safety were the performance standards and evidence was found to show that both the number of accidents decreased and quality improved significantly. In financial terms this revision was a great success.

Incentive schemes which are speed orientated are not usually financially viable when examined closely as part of the safety auditing process. Ones which are safety and quality orientated have been found to be the most cost effective and should be encouraged.

WRITING THE SAFETY AUDIT REPORT

There should be two parts to the safety audit report. These are:

- The executive summary.
- The main body of the report.

Executive summary

This should be similar to an extended abstract and should provide a succinct summary of the methodology, main findings, conclusions and recommendations. The purpose of the executive summary is to provide busy senior managers with as brief a report as possible on what problems were encountered, what the implications are if action is not taken and what decisions need to be taken in order to rectify the problem(s). Ideally, executive summaries should not be longer than ten full sheets of A4 text excluding diagrams and graphs. However, an executive summary will usually be written after a draft of the main body of text has been compiled. Such a précis will follow a logical format such as:

- Background and introduction to the safety audit.
- Methodology employed.
- Findings.
- Conclusions.
- Recommendations.

It is important not to write volumes of text and expect very busy executives to read it all. They will not, they do not have the time. A summary of the salient points is all that is required together with a summary of the decisions that they must take. The main body of text with the details discussed outlined in far more detail will be the property of the safety manager who will be responsible not only for ensuring that where executives require supplementary or additional information they receive it, but also that the safety manager can monitor the action plan implementation programme as recommended by the safety auditors.

Main body of the safety audit report

Volume two will be the part of the report which provides all the background information on which the executive summary was based. It will be necessary in the main report to include all calculations and formulae used and describe every aspect of the safety audit in detail. It will follow a logical format such as:

- A summary of the previous safety audit.
- A summary of the preliminary work carried out by the safety audit team.
- A description of the methodologies employed throughout the safety audit.
- A description of the sequencing which formed the basis of the progress of the safety audit within the organisation.
- A detailed description of the information gathered.
- A detailed description of the analysis of the data.
- A detailed outline of the main findings based on these data.
- A summary of the main conclusions.
- Recommendations.
- An action plan of remedial works and strategies.

The action plan will need to consider:

- the time scale for implementation;
- setting of objectives and goals;
- setting of performance standards and appraisals.

Where tables are presented it would be expected to provide a visual representation of this data for each data set.

Finally, it will be necessary to discuss the main findings of the safety audit with senior managers and this will usually take the form of a presentation. Good overhead projector, slide or computer graphical presentations will be required. It would be normal practice to liaise with the safety manager throughout the safety auditing and report writing stage. Where a safety manager is not employed then the manager who is charged with the responsibility for health and safety matters in the workplace should be involved.

BIBLIOGRAPHY

Hammer, W., *Occupational Safety Management and Engineering*, Prentice-Hall, 1976

Department of Transport, *Accident Investigation Manuals vols 1 and 2*, RoSPA, 1986

Morris, C., *Quantitive Approaches in Business Studies*, Pitman Publishing, 1989

British Safety Council, *Safety Audit check list*, British Safety Council, 1979

Lees, F. P., *Loss Prevention in the Process Industries*, Butterworth and Co, Ltd, 1980

Collins, C. H., *Safety in Clinical and Biomedical Laboratories*, Chapman and Hall, London, 1988

Ferry, T., *Modern Accident Investigation and Analysis*, Wiley Interscience Publications, 1981

HASTAM, CHASE, Health and Safety Technology and Management Ltd.

Rowe, G., Setting Safety Priorities – A Technical and Social Process, *Journal of Occupational Accidents*, 12:31–40

Ferry, T., *Modern Accident Investigation and Analysis*, 2nd Ed, Wiley, 1988

Farmer, D., *Classic Accidents: An Insight into Common Work Accidents*, Croner Publications, 1990

Health and Safety Executive, *Successful Health and Safety Management*, HMSO, 1992

Saunders R. & Wheeler T., *Handbook of Safety Management*, Pitman Publishing, 1991

Saunders R., *Taking Care of Safety*, Personnel Today Series, Pitman Publishing, 1992

INDEX